T0320999

STATIONARY STOCHASTIC PROCESSES FOR SCIENTISTS AND ENGINEERS

STATIONARY STOCHASTIC PROCESSES FOR SCIENTISTS AND ENGINEERS

Georg Lindgren

Lund University
Lund, Sweden

Holger Rootzén

Chalmers University of Technology
Gothenburg, Sweden

Maria Sandsten

Lund University
Lund, Sweden

CRC Press
Taylor & Francis Group
Boca Raton London New York

CRC Press is an imprint of the
Taylor & Francis Group, an **informa** business

A CHAPMAN & HALL BOOK

CRC Press
Taylor & Francis Group
6000 Broken Sound Parkway NW, Suite 300
Boca Raton, FL 33487-2742

© 2014 by Taylor & Francis Group, LLC
CRC Press is an imprint of Taylor & Francis Group, an Informa business

No claim to original U.S. Government works

Version Date: 20130830

International Standard Book Number-13: 978-1-4665-8618-5 (Hardback)

**Visit the Taylor & Francis Web site at
http://www.taylorandfrancis.com**

**and the CRC Press Web site at
http://www.crcpress.com**

Contents

Preface

This text is a first course in stationary stochastic processes for students in science and engineering.

Stochastic processes are indispensable tools for development and research in Signal and Image Processing, Automatic Control, Oceanography, Structural Reliability, Environmetrics, Climatology, Econometrics, and many other parts of science and engineering. Our goal is to help students from these areas learn how to use this tool efficiently. To achieve this requires, we believe, that students are exposed to real-life situations where stationary stochastic processes play a crucial role. It also requires that they acquire sufficient understanding of the theory and enough experience from working with it. Hence, we throughout use problems from science and engineering as a testing ground, and try to strike a good balance between mathematical rigor and ease of exposition. We have laid special emphasis on the interpretation of the various statistical models and concepts and on what questions a statistical analysis can help to answer.

The material (outlined in the next section) in the book is suitable for a one-semester course. The presentation assumes that students already know mathematics corresponding to standard first year university courses in calculus and linear algebra, and that they have some experience with elementary probability and statistics, as typically learnt from introductory statistics courses.

The first version of this text was produced in the early 1970s for a course for electrical engineers at Lund University. At that time focus was on continuous time signal processing, and the content was mainly covariance methods and spectral and Fourier analysis. Since then it has been continually used in courses both for electrical engineers and for students from many other areas, in several of the Swedish universities. It has been updated and improved using the comments of many generations of students and teachers. The exposition has become much broader, reflecting the large increase of the number of areas where stationary stochastic processes are finding significant use. In particular, discrete time methods and time series is a much larger part of the text, and applications come from many more parts of science and engineering. The viewpoint now is that we try to convey broad understanding of the mechanisms which generate stationary stochastic processes, and how to use this understanding in development and research. Thus our aim is to provide generic skills, both for direct use and as the starting point for further specialization, say in signal processing, or time series analysis, or econometrics, or . . . .

Still, Fourier methods remain a main tool, and input-output relations in linear filters continue to be a focal point. In particular, the book contains separate chapters on discrete time auto-regressive and moving average processes, and on continuous time stochastic processes obtained as solutions to linear differential equations. Estimation of mean value functions and covariance functions are treated early in the text, while frequency analysis and estimation of spectral densities have a later chapter of their own.

Many colleagues and students have helped to shape the course, and we are very grateful to them all, but we mention just one – Sven Spanne – who set the style on how to combine mathematical ideas with examples from the "real world."

Georg Lindgren   Holger Rootzén   Maria Sandsten
Lund   Gothenburg   Lund

A survey of the contents

Chapter 1 starts by giving many examples from science and engineering where an observed irregular function can be modeled as a stochastic process, and where this model plays an important role. We then introduce one way to think of a stochastic process: numerical values are generated sequentially in time and with each new value chosen according to some law of chance, where the form of this law depends on the earlier values. The formal statistical tools used to describe this are the finite-dimensional statistical distribution functions. An alternative approach is to regard the observed function as drawn at random from a collection of possible outcome functions, according to some probability law. Fortunately, these two approaches to stochastic processes are equivalent.

To analyze the complete family of finite-dimensional distributions is often difficult, and in many situations it is enough to consider just a few characteristic features of the distributions. Chapter 2 deals with two such features. The mean value function gives information about the statistical expectation of the process when it is observed at a fixed time point. The covariance function measures the degree of dependence between process values taken at two different time points. It also gives information about the variability of the process. The main object of this book, the stochastic processes which are stationary, i.e., such that their statistical properties remain unchanged as time goes by, are then defined. Statistical inference for stochastic processes is the final part of this chapter.

Stationary Poisson processes are statistical models for events which occur randomly in time. Chapter 3 summarizes some properties of these processes. Poisson processes are often part of elementary statistics courses. However, they still defend their place in this book, since they are the simplest and most important stationary point process. More general point processes are also mentioned in this chapter.

Chapter 4 is devoted to Fourier analysis of the covariance function. In the context of this book this is usually called spectral analysis. The spectral analysis represents a stationary stochastic process as a sum (or more generally, an integral) of simple cosine function with random amplitudes and phases. This is a central concept in the theory of stationary processes. The relation to Fourier analysis of data is also discussed in this chapter. In particular we describe how sampling of a continuous time process at discrete time points affects the spectrum, and how one can reconstruct a continuous processes from a sampled version.

The normal, or Gaussian, distribution is the most important of all statistical distributions. Chapter 5 deals with the normal stochastic processes, for which all finite-dimensional distributions are normal. These distributions are completely determined by the mean value and covariance functions. The important Wiener process (often instead called a Brownian motion process) is also described in this chapter, as well as the more general Lévy processes.

Chapters 6–8 deal with how stationary processes change when they pass through a linear filter. Differentiation, integration, and summation are examples of such filters. We describe how the mean value and covariance function, as well as the spectrum, are affected by passage through a filter. The general theory is presented in Chapter 6. This chapter also contains a discussion of white noise process. Cross-covariance and cross-spectrum are introduced as measures of the relation between two different processes.

Chapter 7 is devoted to a very useful class of stationary processes, namely the auto-regressive (AR) and moving average (MA) processes. These are obtained by passing white noise in discrete time through a recursive (feedback) or a transversal filter, respectively. These processes are important building blocks in the analysis of economic time series, and in stochastic control theory.

Chapter 8 contains applications of linear filters in continuous time, including differentiation and integration, and defines the concept of an envelope. Simple linear stochastic differential equations are introduced. Examples of optimal filters in signal processing (matched filter and the Wiener filter) are presented, together with the ideas behind the Kalman filter.

Chapter 9 contains the data analytic counterpart of the probabilistic spectral theory in Chapter 4, and studies Fourier analysis of data. First, the raw periodogram is obtained as the Fourier transform of an observed stationary stochastic process. Then it is shown that it is not a good estimator of the spectral density of a stationary process since its variance does not improve with an increasing number of observations. The periodogram hence has to be modified to provide good estimates of the spectrum. Several such modifications are described in this chapter.

Monte Carlo simulation of stochastic processes can serve as a link between theory and practice, and it can help to understand the properties of a stochastic model. We have therefore included in many of the chapters hints on how to generate samples of the described models.

The book concludeds with five appendices: basic probability and statistical theory, including the multivariate normal distribution; delta functions and the Stieltjes integral; the Kolmogorov existence theorem; a table of Fourier transforms; a historic background featuring some of the early development of stationary processes.

Suggestions for reading

The text is intended for "a first course" for students in science and engineering. However, some selection of material is advisable for a basic course.

- For a first course, general audience:
  - Chapter 1
  - Sections 2.1–2.3; 2.4.1–2.4.2; 2.5.1–2.5.2
  - Sections 3.1–3.3; 3.5–3.6
  - Sections 4.1–4.4
  - Sections 5.1–5.2; 5.3.1; 5.3.3
  - Sections 6.1–6.3; 6.5.1
  - Sections 7.1–7.3
- Additional material be chosen depending on specialization;
  - Sections 4.5.1–4.5.2 (The sampling theorem, Fourier inversion)
  - Sections 5.4–5.5 (Chi-squared, Lévy, and shot noise processes)
  - Sections 6.4; 6.5.2 (White noise in continuous time)
  - Sections 7.3; 7.4.1; 7.5 (Estimation and prediction, GARSH processes)
  - Section 8.3 (The envelope)
  - Sections 8.4–8.5 (Matched filter, Wierner filter)
  - Chapter 9 (Spectral Estimation)
  - Sections 2.6, 3.7, 5.6, 7.6 (Monte Carlo simulation)

Chapter 1

Stochastic processes

1.1 Some stochastic models

Stochastic models are used to describe experiments, measurements, or more general phenomena for which the outcomes are more or less random and unpredictable. We use the word "unpredictable" in a very weak sense that it is not possible to exactly predict what the result of the experiment will be. But it does not mean that we lack knowledge about the possible outcomes. In fact, we could have very precise knowledge about the likelihood of different results. This knowledge can be theoretical, based on probabilistic assumptions and arguments, but most often it also includes empirical data. *Estimation* of the *model parameters* is an important part of the modeling process.

The basic elements in a stochastic model are *random variables*, introduced to handle the fact that the result of the experiment may be one or several numerical values, each describing some aspect of the experiment. This vague "definition" of a random variable can be given a more strict mathematical meaning, and we will touch upon this in Section 1.2. However, the notion of a random variable is important in its statistical context, and should not be thought of only as a mathematical concept.

Random variables are often denoted by capital letters, X, Y, Z, etc. The actual numerical results in a single experiment are referred to as *observations* or *realizations* of the random variables, and are denoted by the corresponding lower case letters, x, y, z, etc.

If the result of the experiment is a function of a variable t, for example of time, we talk about a *random function* or a *stochastic process*, and use the notation $\{X(t), t \in T\}$, when t varies in the set T, usually an interval, or a discrete set of time points. In this book we use the term "stochastic process" and reserve the term "random function" for situations when we want to emphasize that a function has been chosen at random.

A formal definition of a stochastic process is given in Section 1.2.

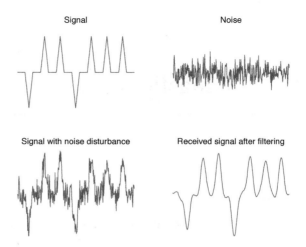

Figure 1.1 *Digital transmission disturbed by random noise.*

We will now present a range of examples of experiments or situations, where it is natural to use a stochastic process to model the randomness.

Example 1.1 ("Digital transmission"). Figure 1.1 shows parts of the transmission of zeros and ones, represented by positive and negative peaks, over a noisy channel. During the transmission, noise is added to the signal, making the received signal quite distorted. A filter, specially designed to detect the positive and negative pulses, removes most of the noise. The noise in Figure 1.1 is an observation or realization of the stochastic noise process, which we can denote by $\{X(t), 0 \leq t \leq T\}$, where T is the length of the recording.   ▲

Example 1.2 ("Speech analysis"). Speech analysis with different applications, such as speech recognition, speech synthesis, and speaker recognition, is an area where stochastic processes are highly important. An example of recorded speech signals is seen in Figure 1.2 where the word 'Hallå' is recorded in two different Swedish dialects, 'Skånska' from the Southern part, and 'Norrländska' from the Northern part.   ▲

The processes in Examples 1.1–1.2 are said to have *continuous parameter* t, or, more commonly, *continuous time*. The realizations are functions of a real variable t, which takes values in an interval T. A stochastic process can also be a sequence of random variables, and then it is said to have *discrete parameter* or *discrete time*. Processes with discrete time are often called *time series* and denoted $\{X_t\}$. If a continuous time process is *sampled* at discrete,

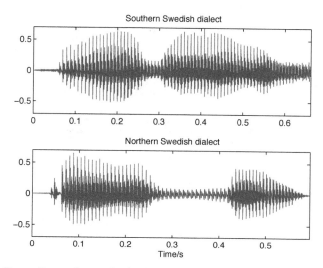

Figure 1.2 *Recordings of speech, the Swedish word 'hallå' in two different dialects.*

equidistant times, one obtains a random sequence. Such are studied in detail in Sections 4.3 and 4.4.

Example 1.3 ("Automatic control"). An autopilot on board a ship intends to keep the ship on a fixed course. Figure 1.3 shows a result from an early experiment with stochastic control. The curve shows the deviation between the intended and the real heading during a half hour experiment with stochastic steering control on board the Wallenius/Atlantic Container Line 15.000-ton container ship Atlantic Song. (Data from Åström et al.: The identification of linear ship steering dynamics, [59].) The sampling interval is 15 seconds.

If X_k denotes the deviation at time $t = 15k$ (measured from the beginning of the experiment) we can model X_k as a sequence of dependent random variables $\{X_k, k = 1, 2, \ldots, 120\}$, i.e., a stochastic process with discrete "time." The figure shows a realization of the process. A new experiment will give a different result. ▲

In all the presented examples, the *range* or *state space* has been continuous, i.e., the set of possible process values has been a real interval or even the entire real line $\mathbb{R}$. A range can also be discrete, e.g., when the process takes only integer values.

Example 1.4 ("Telephone traffic"). Telephone traffic, now more than a century old, was one of the first industrial applications of stochastic process theory, and it generated the first serious study of the *Poisson process*.

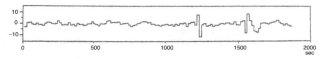

Figure 1.3 *Container ship Atlantic Song (photo: Wallenius Lines) and deviation between intended and real heading in autopilot experiment (from [59]).*

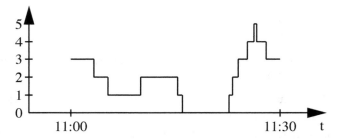

Figure 1.4 *Number of outgoing phone calls from a base station.*

The number of active connections in a part of a telephone network varies in an irregular way, and can be modeled by stochastic processes: e.g., if $X(t)$ is the number of outgoing phone calls from a base station at time t, then $X(t)$ is the difference between $Y(t)$, the number of calls that started before time t, and $Z(t)$, the number of calls that ended before t. As functions of t, both $Y(t)$ and $Z(t)$ are stochastic processes with range $\mathbb{N} = \{0, 1, 2, \ldots\}$. ▲

Example 1.5 ("Stock market prices"). Prices on the stock market vary, systematically and randomly. Long term trends and regular business cycle variations are combined with seemingly random fluctuations. When the systematic and seasonal effects have been taken care of, the statistical laws of the residuals are often quite stable, at least for some time, and they can therefore be modeled by a stationary stochastic process. Abrupt changes in the behavior occur, however; much of the theory for non-linear time series, such as GARCH-models, deals with such changes in the mechanism that underlie the

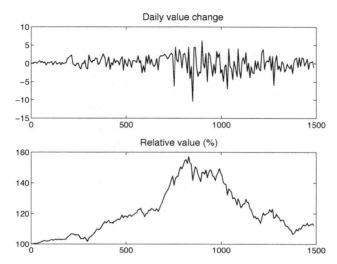

Figure 1.5 *Value changes in a trust fund during 1500 days. Upper diagram shows the daily changes, and the lower diagram illustrates the relative value from the start.*

fluctuations. Figure 1.5 shows the development of the value in a Swedish trust fund during 1500 days. In another direction, the Wiener process is a central ingredient in the Black-Scholes-Merton model for option pricing and hedging. This model has had an enormous impact on the functioning of the world's financial markets.   ▲

Example 1.6 ("Stochastic fatigue"). Cars driving on an uneven road, ships that load, unload, and cross the ocean, airplanes which take off, climb, descend, and land are all subject to random oscillating stresses. These stresses can cause slowly growing cracks in the components of the vehicle. After a long time these fatigue cracks can lead to the component breaking into pieces, sometimes with catastrophic consequences.

The oscillating stresses are caused by inhomogeneities in the environment, waves, air turbulence, etc., and the vehicle acts as a linear (or nonlinear) filter. In Section 8.7, we will use stationary stochastic processes models to analyze these oscillations in the simplified setup illustrated in Figure 1.7.

More advanced models of this kind are important ingredients in the construction of safe and reliable vehicles.   ▲

Example 1.7 ("Water quality"). Water quality in rivers is regularly monitored by measurements of pH, oxygen and oxygen demand, and natural or anthropogenic substances. Phosphorus (P) is an element that mainly origi-

Figure 1.6 *Heavy Scania truck on rough road in Indonesia and on a randomized test rig; photo: Carl-Erik Andersson, Scania AB.*

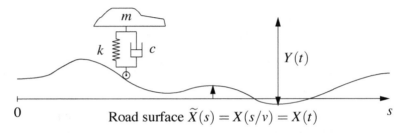

Figure 1.7 *A simple one-dimensional model of a road and a car (for notation, see Section 8.7.1).*

nates from the public sewage system and agriculture, whereas silicon (Si) is mainly of natural origin, by decomposition of natural minerals. Decomposition also contributes with a small proportion of the phosphorus concentration.

Figure 1.8 shows monthly measurements of phosphorus (upper curve) and silicon (lower curve) [mg/l] in the Swedish river Ljungbyån near Ljungbyholm, from the years 1965–2000. The phosphorus series is very irregular, but one can clearly see an effect of the new technique that was introduced in the sewage treatment plant in Nybro in August 1973. The question is now: How efficient was this new step in the reduction of phosphorus? The answer should be given as a *confidence interval*, (P_1, P_2), that with 95% confidence covers the true (average) reduction.

Here is a problem: the monthly measurements are not independent but rather positively correlated. This leads to increased uncertainty in the estimation of the average reduction, compared to independent data. To account for this increase one can use some method from statistical *time series analysis*, which is based on the stochastic process theory presented in this book. In fact, already in 1986 one could claim that with 95% confidence, the phos-

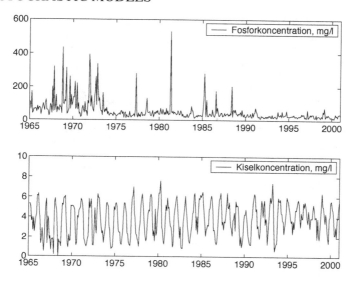

Figure 1.8 *Phosphorus (upper) and silicon (lower) concentration in Ljungbyån.*

phorus reduction was between 40% and 80%. (The example is taken from L. Zetterqvist: Statistical methods for analysing time series of water quality data, [57].) The silicon concentration has a different variation pattern, since it is strongly dependent on the season of the year. Given the covariation between silicon and the phosphorus that is of natural origin, one can hope for even more precise estimates of the sewage phosphorus. ▲

In the examples 1.1–1.7, the parameter t was one-dimensional, and it often represented "time." There are also many important examples of random functions of two or more variables.

Example 1.8 ("Image analysis"). A digital image consists of black or colored dots, pixels. The "parameter" t is the pixel coordinates, varying over the image area T, and the intensity at pixel t is $X(t)$. The content of the image can be quite diverse, and a stochastic model for the content is often appropriate. The type of model can be chosen to suit the application, but within the chosen model one strives to allow for correct interpretation of the model, at the same time as one is prepared to allow also for unexpected elements.

When an image is broadcasted, noise can cause salt and pepper disturbances, consisting of black or white random dots on the surface. These can often be removed by a simple manipulation, and the true image almost restored. Figure 1.9 shows an old view from the Mathematics building in Lund towards the old water tower. The left image has been artificially disturbed by randomly making the pixels either black or white with probability 0.05.

Figure 1.9 *Noisy photo of the old water tower in Lund; to the right the photo is cleaned by a five point median filter.*

The right image is cleaned by a five point median filter to the disturbed image. This means that each pixel has been colored according to a majority vote among its four neighbors.   ▲

Example 1.9 ("Astronomy"). Stochastic process models for the spatial distribution of radiation, matter, stars, and galaxies in an expanding universe, and statistical methods for these models are standard ingredients in astronomy.

One example is the *Scott effect*, named after the statistician Elizabeth Scott: it is the observational bias of very distant clusters of galaxies. The more galaxies the clusters contain, the more likely it is that they are observed, and hence a bias correction is needed. Figure 1.10 shows one of the biggest and most distant clusters of galaxies observed so far, as an example of a two-dimensional random field generated by a three-dimensional random space distribution.

Anisotropy, small deviations from stationarity, in the cosmic background radiation can confirm or reject the Big Bang theory of the universe. Fourier

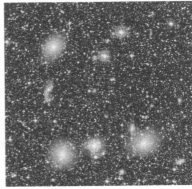

Figure 1.10 *A galaxy cluster in the universe as a spatial stochastic process; ESO, European Southern Observatory, 1999.*

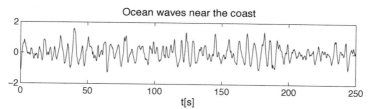

Figure 1.11 *Wave recordings at a wave pole near the West African coast.*

analysis methods for stationary random fields are hence crucial for cosmology. ▲

Example 1.10 ("Ocean waves"). Random wave models have been used in ocean engineering since about 1950 in ship design and for the study of ship response to waves. Figure 1.11 gives a recording of waves as a time function, measured at a fixed wave pole.

The sea surface is a good example of a stochastic process with multi-dimensional parameter. The coordinates on the surface are a two-dimensional space parameter and time is a third parameter; the model is spatio-temporal. Let the height of the sea surface at time t at the point $\mathbf{u} = (u_1, u_2)$ be denoted by $X(t, \mathbf{u})$. With fixed $\mathbf{u}$ and varying t, we get the process of sea level variation at point $\mathbf{u}$.

Keeping t fixed with varying $\mathbf{u}$ gives an image of the sea surface at time t, also called a *random field*. Figure 1.12 shows a photo of ocean waves in the open sea. The winds had been steady for some time and the waves are

Figure 1.12 *Top: Photo of the sea surface during stable weather; (Earth Sciences Australia). Bottom: Randomly simulated sea surfaces with directional spreading.*

Figure 1.13 *Random Santas.*

long-crested, going mainly in one single direction. The figure also shows two artificially generated surfaces, $X(t_{\text{fix}}, \mathbf{u})$, with more irregular waves. ▲

Example 1.11 ("Christmas card"). Let us finish with an example to show that the elements of a stochastic process can also be more general objects. In Figure 1.13, randomly sized Santa Claus figures are randomly spread out over a Christmas card from the early nineteen seventies. Each pixel carries information about the presence or absence of a figure, including its size and color. This is an example of a *marked point process*, where the point locations form a random point process, and to each point there is attached a mark, that can also be random; in this case the marks are size and color. ▲

1.2 Definition of a stochastic process

The term stochastic process is a rather formal name for an everyday phenomenon. A more prosaic name would be *random function*, a name that hints at its basic property: a function is chosen at random from a set of possible functions, where the choice is governed by some statistical law. The chosen function, the realization, is a function of some time variable t, usually real. Thus, there are two types of variation involved: the variation *between* different realizations, and the variation with time *within* the chosen function. In all the examples we have presented so far, the realizations have looked quite irregular and "random." Note, however, that the term random function refers to the choice of functions, not the appearance; see Example 1.12, in which all sample functions are very regular cosine functions. Only the phase and amplitude are chosen at random.

We are now ready to give a mathematical definition of a stochastic process. First recall the notion of *sample space*, Ω, for an experiment, defined as any set such that every possible outcome of the experiment is represented by exactly one element ω in the set. (Some elements may not correspond to any possible outcome, but it is important that everything that really can happen is there.) A random variable on Ω is then defined as a real valued function $\omega \mapsto X = X(\omega)$, with certain properties, defined on Ω.

Definition 1.1. *A stochastic process with parameter space T is a family*

$$\{X(t), t \in T\}$$

of random variables, defined on a sample space Ω.

If T is a real line, or a part of the real line, the process is said to have continuous time. If T is a sequence of integers, the process is said to have discrete time, and it is called a random sequence, or time series. If T is a set in space, or space and time, the process is called a random field (Since the value of the process depends both on the outcome ω of the random experiment and on t, a more complete notation for the function values would be $X(t, \omega)$. However, as is customary in probability theory, we usually suppresses the argument ω.)

For fixed outcome ω, the function $t \mapsto X(t, \omega)$ is, as mentioned, called a *realization* of the process. Other names are *sample path*, *trajectory*, and *sample function*. The set of possible sample paths is called the *ensemble* of the process.[1]

Processes with continuous time will most commonly be of the type $\{X(t), t \in \mathbb{R}\}$ or $\{X(t), t \geq 0\}$. Processes with discrete time are most often of the type $\{X(t), t = \ldots, -1, 0, 1, 2, \ldots\}$ or $\{X(t), t = 0, 1, \ldots\}$, and are called random sequences or time series.

In this text, we will frequently suppress the notation "$t \in T$" in a stochastic

---

[1] The ensemble for a stochastic process consists of those functions that can occur as the result of a certain experiment. In some situations one can use the ensemble set as sample space Ω. Then $\omega \in \Omega$ is the function that represents the outcome of the experiment and $X(t, \omega) = \omega(t)$.

However, in general, the sample space is larger than the ensemble, for example when one measures two different data sets in the experiment. As a simple example, think of throwing one coin and one die several times, defining $x_t = 1$ if the coin shows "head" in throw number t, and zero otherwise, and with y_t equal to the die outcome. An outcome in the full sample space could be a sequence of pairs with 0/1 as first element and 1/2/3/4/5/6 as second element. The ensemble for the coin outcomes is a sequence of zeros and ones.

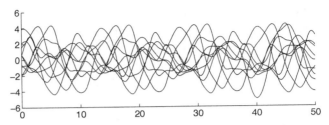

Figure 1.14 *Ten different realizations of the same stochastic process.*

process, and write only $\{X(t)\}$ or $\{X_t\}$. However, $X(t)$, without brackets, is always the value of the process at time t, and hence a random variable.

Figure 1.14 shows ten different realizations of a stochastic process. They are all different, but have rather similar appearances since they are generated by the same random mechanism.

We now come to an extremely important, although simple, example of a stochastic process, the cosine process. Such processes play an important role for one of the main themes of this book, Fourier analysis of stationary processes.

Example 1.12 ("Random phase and amplitude"). Let A and ϕ be two random variables, such that $A > 0$ and $0 \leq \phi < 2\pi$, and define a cosine function with A as "amplitude" and ϕ as "phase,"

$$X(t) = A\cos(t + \phi). \tag{1.1}$$

Note that a single realization of A, ϕ produces a realization of $X(t)$ for all t.

For each time t, we have now defined a random variable $X(t)$. The sequence $\{X(t), t = 0, 1, \ldots\}$ is a stochastic process with discrete time, and $\{X(t), t \in \mathbb{R}\}$ is a process with continuous time. The random choice of amplitude and phase gives an outcome (A, ϕ), and as sample space we can take $\Omega = \{(x, y), x \geq 0, 0 \leq y < 2\pi\}$. As an alternative, we could as sample space choose the set of all continuous functions. A more reasonable sample space would be the set of cosine functions with arbitrary phase and amplitude. The realizations are plain cosine functions, which do not look very random! There is no way to determine, from a single realization, that we have to deal with a stochastic process.

In the sequel we will usually assume that A and ϕ have continuous distribution. However we will now use a very much simplified example to illustrate how outcomes $\omega = (A, \phi)$ are mapped to realizations of the stochastic process. In Figure 1.15, we have picked two possible outcomes, $(A, \phi) = (2, 3\pi/2)$, and $(A, \phi) = (1, \pi)$, and plotted the corresponding functions. If we

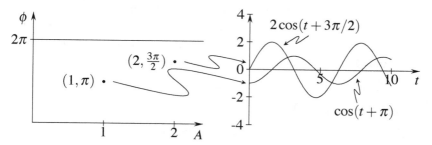

Figure 1.15 *Sample space (left) and ensemble space (right) for the process $X(t) = A\cos(t + \phi)$ for two different outcomes of (A, ϕ).*

assign probabilities just to these two points, for example $1/3$ and $2/3$, we have constructed an extremely simple example of a random process: just select one of the points with the correct probability, and draw the corresponding function graph. The resulting cosine function is the realization. Thus, a realization of a stochastic process need not be irregular! The important property is that it is the result of a random experiment.

In this example we use the sample space Ω of (A, ϕ) and the ensemble of cosine function outcomes in quite different ways. It is easy to assign probabilities to the points in the sample space by use of elementary probability while it requires more abstract thinking to assign probabilities to the members in the ensemble of outcome functions.   ▲

This example may look artificial, because there does not seem to be much "randomness" in a single cosine function. However, it is our first encounter with one of the most important building blocks in the theory of stationary stochastic processes. By adding together an increasing number of cosine functions, $A_k \cos(2\pi f_k t + \phi_k)$, all with different amplitudes and phases, and also with different frequencies f_k, the sum starts to look more and more "random" and unpredictable; see Figure 2.13 for an illustration of randomness from just three terms.

A sum of many, even infinitely many, random cosine functions will be a recurrent example when we analyze and characterize the statistical properties of stationary processes, and their applications. Signal processing by means of Fourier analysis and the treatment of random signals in linear filters are the topics of Chapters 4, 6–9.

Remark 1.1. *Instead of the formulation in Definition 1.1, we could have defined a stochastic process as a randomly chosen function, or more precisely as a random variable with values in a function space. This requires that one*

*can define probabilities for sets of functions, for example the set of contin-
uous functions. However, this is somewhat complicated mathematically. In
Appendix C we will touch upon this problem, when we discuss the basic theorem for stochastic processes, namely Kolmogorov's existence theorem.*

Summary of notation and concepts:

Ω	= sample space for a random experiment
T	= parameter space, e.g.,
	$T = \{0, 1, \ldots\}$ or $T = \mathbb{R}$
$X(t), X_t$	= single random variable defined on Ω
$\{X(t), t \in T\}, \{X(t)\}$	= family of random variables,
	stochastic process
$x(\cdot), x$	= $\begin{cases} \text{function } t \mapsto x(t) \text{ (outcome } \omega \text{ fixed)} \\ \text{sample path or realization} \end{cases}$
$\{x(t), t \in T\}$	= ensemble of outcome functions or
$\{x_t, t \in \mathbb{Z}\}$	sequences

1.3   Distribution functions

Any random variable X has a *cumulative distribution function (cdf)*,

$$F_X(x) = \mathsf{P}(X \leq x),$$

that gives the probability of the event that X takes a value less than or equal
to x. The variable can be *continuous*, with *probability density function (pdf)*

$$f_X(x) = \frac{dF_X(x)}{dx}$$

or *discrete*, with *probability function* $p_X(x) = \mathsf{P}(X = x)$. The cdf determines
all statistical properties of X.

For two random variables, the bivariate distribution function is
$F_{XY}(x, y) = \mathsf{P}(X \leq x, Y \leq y)$, with bivariate density function $f_{XY}(x, y) =
\frac{\partial^2 F_{XY}(x,y)}{\partial x \partial y}$, and probability function $p_{XY}(x, y) = \mathsf{P}(X = x, Y = y)$. Two random
variables are called *statistically independent* if

$$F_{XY}(x, y) = F_X(x) F_Y(y),$$

with the corresponding product formulas for the density or probability functions. We remind the reader of the definitions of expectation and variance

for a random variable, and the covariance and correlation coefficient between two random variables. Further, recall that if ρ is the correlation coefficient between two random variables X and Y, then $-1 \le \rho \le 1$, and if $\rho = 1$ or -1 then X and Y are linear functions of one another.

Expectation: $E[X] = m_X = \begin{cases} \int x f_X(x)\,dx, \\ \sum x p_X(x), \end{cases}$

Variance: $V[X] = E[(X - m_X)^2] = E[X^2] - m_X^2,$

Standard deviation: $D[X] = \sqrt{V[X]},$

Covariance: $C[X,Y] = E[(X - m_X)(Y - m_Y)] = E[XY] - m_X m_Y,$

Correlation coefficient: $\rho_{X,Y} = \dfrac{C[X,Y]}{\sqrt{V[X]V[Y]}}, \quad |\rho_{X,Y}| \le 1.$

If $C[X,Y] = 0$, the variables X and Y are said to be uncorrelated. Two independent variables are always uncorrelated, but two uncorrelated variables need not be independent.

More details on how to handle distribution function, in particular how to use delta functions to describe mixtures of discrete and continuous distribution functions, are given in Appendix B.

1.3.1 The CDF family

If $\{X(t), t \in T\}$ is a stochastic process, then $X(t)$ is a random variable and hence it has a distribution function, $F_{X(t)}$,

$$F_{X(t)}(x) = P(X(t) \le x),$$

which gives the probability that the random function takes a value less than, or equal to, the real number x at time t. Instead of $F_{X(t)}$, we will write F_t when there is no doubt about which process it refers to. The distribution $F_t(x)$ is called the *marginal distribution* of $X(t)$.

In the same way $(X(t_1), X(t_2))$ is a two-dimensional random variable with distribution function F_{t_1,t_2},

$$F_{t_1,t_2}(x_1, x_2) = P(X(t_1) \le x_1, X(t_2) \le x_2),$$

meaning that the random function takes a value less than or equal to x_1 at time t_1, and, in the same realization, it is less than or equal to x_2 at time t_2. This generalizes to the n-dimensional distribution function $F_{t_1,\dots,t_n}$,

$$F_{t_1,\dots,t_n}(x_1, \dots, x_n) = P(X(t_1) \le x_1, \dots, X(t_n) \le x_n).$$

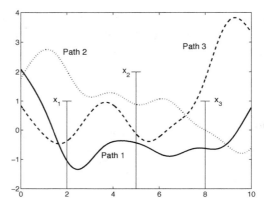

Figure 1.16 *Conditions on the random function $X(t)$ at three time points: $X(t_k) \leq x_k, k = 1, 2, 3$. Of the three realizations, only path number 1 fulfills the conditions.*

The n-dimensional distribution restricts the sample functions in many ways. In Figure 1.16, we have illustrated this for three different realizations ("paths") of a stochastic process $X(t)$, with the three constraints at times $t_1 = 2$, $t_2 = 5$, and $t_3 = 8$, that it should be less than or equal to 1, 2, and 1, respectively. Only path number 1 satisfies all three conditions.

Definition 1.2. *The family F_T of distribution functions,*

$$
\begin{aligned}
F_t \quad &; \quad t \in T \\
F_{t_1, t_2} \quad &; \quad t_1, t_2 \in T \\
\vdots \quad &\qquad \vdots \\
F_{t_1, \dots, t_n} \quad &; \quad t_1, \dots, t_n \in T \\
\vdots \quad &\qquad \vdots
\end{aligned}
$$

is called the family of finite-dimensional distributions, or more briefly, the F-family for the stochastic process $\{X(t), t \in T\}$.

Remark 1.2 (The "distribution" of a stochastic process). *By the distribution of a stochastic process, we usually mean the family F_T of finite-dimensional distributions. In general it can be quite difficult to specify the entire F_T-family for a stochastic process. However, for some very important cases, such as normal processes and Poisson processes, this is simpler, as will be seen in Chapters 3 and 5.*

1.3.2 Gaussian processes

Gaussian processes are the most important stationary processes. In Chapter 5, we will consider such processes in detail, and also make use of the multi-dimensional normal distribution. Appendix A.2 deals with the multidimensional normal distribution.

Example 1.13 ("Noise"). If $X(t)$ is the random voltage fluctuations caused by the noise which is added to a signal when it is transmitted over a radio channel it might often be reasonable to assume that $X(t)$ is normally distributed with mean zero and some standard deviation σ, i.e., $X(t) \sim N(0, \sigma^2)$, at least approximately. However, just knowing the mean and variance does not tell us much about the noisiness of the signal; it only defines the proportion of values between specified limits, for example that most of the values (about 95%) will be between $\pm 2\sigma$.

The two-dimensional distributions can say something about the smoothness of the output. For instance, if $X(t_1)$ and $X(t_2)$ are almost independent for all t_1 and t_2, not too close to each other, then the voltage will fluctuate quickly, as in the upper plot in Figure 1.17. If, on the other hand, there is a strong positive correlation between neighboring $X(t_1)$ and $X(t_2)$, the variations will be slower, but of the same magnitude, as in the lower plot. ▲

The distribution of $X(t)$ for single t-values alone does not specify much of the statistical properties of a stochastic process, as we have seen in Example 1.13. One also needs the higher dimensional distributions to link several $X(t_1), X(t_2), \ldots, X(t_n)$ to each other.

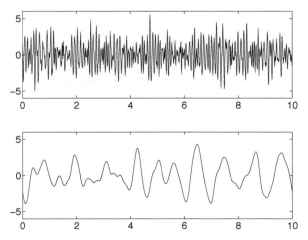

Figure 1.17 *Noise with small (top) and large (bottom) positive temporal dependence.*

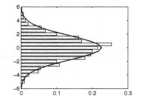 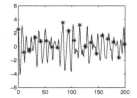

Figure 1.18 *Histogram with normal density (left) for wave measurements (right).*

Example 1.14 ("Ocean waves"). Ocean waves, described in Example 1.10, can be nice examples of a Gaussian process. The right plot in Figure 1.18 shows variations in the sea surface height at a measuring station, sampled every second; to the left is a histogram over every 10th value, marked with stars in the diagram to the right. The total observation time is three hours, and the histogram is based on 1000 values. A normal density with mean zero and standard deviation estimated from the data is overlaid the histogram.

Figure 1.19 illustrates the covariation of measurements with two different time separations. Each dot represents a pair $(x(t), x(t+h))$, with $h = 5$ and $h = 10$ seconds, respectively. The value on the vertical axis is thus taken h seconds after the value on the horizontal axis. Note that with 5 seconds separation, positive and negative values seem to alternate, while for the double separation, values have more often the same sign. One could think of this as a random function with a tendency of a 10 second period.   ▲

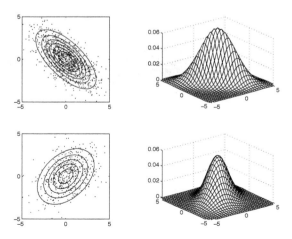

Figure 1.19 *Scatter plots and two-dimensional normal densities for pairwise wave measurements with different time separation: 5 seconds (top), 10 seconds (bottom).*

Chapter 2

Stationary processes

2.1 Introduction

In this chapter we introduce two basic tools for analysis of stationary processes: the mean value function, which is the expected value of the process as a function of time t, and the covariance function, which is the covariance between process values at times s and t. We review some simple rules for expectations and covariances, for example, that expectations are linear, and covariances are linear in both arguments. We also give many examples of how the mean value and covariance functions can be interpreted. The main focus is on processes for which the statistical properties do not change with time – they are (statistically) stationary. Strict stationarity and weak stationarity are defined.

The study of a stochastic process as a function of time is called analysis "in the time domain." To prepare for analysis "in the frequency domain," which is the topic of Chapter 4, we describe a simple type of stationary process, namely, the sum of cosine functions with random amplitudes and phases.

The statistical problem of how to find good models for a random phenomenon is also dealt with in this chapter, and in particular how one should estimate the mean value function and covariance function from data. The dependence between different process values needs to be taken into account when constructing confidence intervals and testing hypotheses.

A section on Monte Carlo simulation of stationary sequences from the covariance function concludes the chapter.

2.2 Moment functions

The statistical properties of a stochastic process $\{X(t), t \in T\}$ are determined by the F_T-family of finite-dimensional distribution functions. Expectation and standard deviation catch two important properties of the marginal distribution of $X(t)$, and for a stochastic process these may be functions of time. To describe the time dynamics of the sample functions, we also need some simple

19

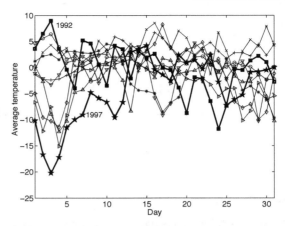

Figure 2.1 *Daily average temperature in Målilla during January, for 1988–1997. The fat curves mark the years 1992 and 1997. Data: Carl-Erik Fröberg.*

measures of the dependence over time. The statistical definitions are simple, but the practical interpretation can be complicated. We illustrate this by the simple concepts of "average temperature" and "day-to-day" correlation.

Example 2.1 ("Daily temperature"). Figure 2.1 shows the daily average temperature in the Swedish village of Målilla during the month of January for the ten years 1988–1997. Obviously, there have been large variations between years, and it has been rather cold for several days in a row. This variation between years is obviously an important fact to be considered when one wants to assess climate change.

The atmospheric circulation is known to be a very chaotic system, and it is hard to predict the weather more than a few days ahead. However, modern weather forecasts adopt a statistical approach in the predictions, together with the computer intense numerical methods, which form the basis for all weather forecasts. Nature is regarded as a stochastic weather generator, where the distributions depend on geographical location, time of the year, etc., and with strong dependence from day to day. One can very well imagine that the data in the figure are the results of such a "weather roulette," which for each year decides on the dominant weather systems, and on the day-to-day variation. A changing climate may be thought of as slowly changing odds in the roulette.

With the statistical approach, we can think of the ten years of data as ten observations of a stochastic sequence $X_1, \ldots, X_{31}$. The *mean value function* is $m(t) = \mathsf{E}[X_t]$. Since there is no prior reason to assume any particular values for the expected temperatures, one has to rely on historical data. In meteorology, the observed mean temperature during a 30 year period is used as standard.

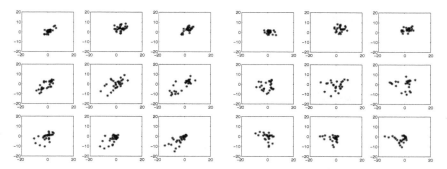

Figure 2.2 *Scatter plots of temperatures from nine years, 1988–1996, for two successive days (left plot) and two days, five days apart (right plot). One can see a weak similarity between temperatures for adjacent days, but it is hard to see any connection with five days separation.*

The covariance in the temperature series can also be seen in the data. Figure 2.2 illustrates the dependence between temperatures. For each of nine years we show to the left scatter plots of the pairs (x_t, x_{t+1}), with temperature one day on the horizontal axis and temperature the next day on the vertical axis. There seems to be a weak dependence; two successive days are correlated. To the right we have similar plots, but now with five days separation, i.e., (x_t, x_{t+5}). There is almost no correlation between days that are five days apart. ▲

2.2.1   The moment functions

We now introduce the basic statistical measures of average and correlation. Let $\{X(t), t \in T\}$ be a real stochastic process with discrete or continuous time.

Definition 2.1. *For any stochastic process, the first and second order moment functions are defined as*

$$
\begin{aligned}
m(t) &= \mathsf{E}[X(t)] & \text{mean value function} & \quad (mvf) \\
v(t) &= \mathsf{V}[X(t)] & \text{variance function} & \quad (vf) \\
r(s,t) &= \mathsf{C}[X(s), X(t)] & \text{covariance function} & \quad (cvf) \\
b(s,t) &= \mathsf{E}[X(s)X(t)] & \text{second-moment function} & \\
\rho(s,t) &= \rho[X(s), X(t)] & \text{correlation function} &
\end{aligned}
$$

There are some simple relations between these functions:

$$r(t,t) = \mathsf{C}[X(t),X(t)] = \mathsf{V}[X(t)] = v(t),$$
$$r(s,t) = b(s,t) - m(s)m(t),$$
$$\rho(s,t) = \frac{\mathsf{C}[X(s),X(t)]}{\sqrt{\mathsf{V}[X(s)]\mathsf{V}[X(t)]}} = \frac{r(s,t)}{\sqrt{r(s,s)\,r(t,t)}}.$$

Remark 2.1. *We have defined the second-moment function as* $b(s,t) =$ $E[X(s)X(t)]$. *In signal processing literature,* $b(s,t)$ *is often called the "auto-correlation function." Since in statistics, the technical term correlation is reserved for the normalized covariance, we will not use that terminology. In daily language, correlation means just "co-variation." A practical drawback with the second moment function is that the definition does not correct for a non-zero mean value. The covariance function is equal to the second moment function for the mean value corrected series* $X(t) - m(t)$.

The moment functions provide essential information about the process. The meaning of the mean value and variance functions are intuitively clear and easy to understand. The mean value function describes how the expected value changes with time; for example, we expect colder weather during winter months than during summer. The (square root of the) variance function tells us what magnitude of fluctuations we can expect.

Covariances measure the amount of linear dependence. For example, in the ocean wave example, Example 1.14, the covariance $r(s,s+5)$ is negative and $r(s,s+10)$ is positive, indicating that measurements five seconds apart often fall on the opposite side of the mean level, while values at ten seconds distance often are on the same side. The covariance is a *linear* measure of the similarity between observations taken at different time points.

If there is more than one stochastic process of interest, say $X(t)$ and $Y(t)$, one can distinguish their moment functions by indexing them, as m_X, r_X and m_Y, r_Y, etc. A complete name for the covariance function is then *auto-covariance function*, to distinguish it from a *cross-covariance function* which measures the (linear) dependence between two different stochastic processes. In Chapter 6, we will investigate this measure of co-variation further.

Definition 2.2. *The function,*

$$r_{X,Y}(s,t) = \mathsf{C}[X(s),Y(t)] = \mathsf{E}[X(s)Y(t)] - m_X(s)m_Y(t),$$

is the cross-covariance function between $\{X(t)\}$ *and* $\{Y(t)\}$.

If two random variables are independent then they are also uncorrelated. However, we again emphasize that correlations only measure linear dependence: Uncorrelated variables can very well have a strong (non-linear) dependence, or even be completely dependent, as seen in the next example.

Example 2.2 ("Two uncorrelated random variables which are completely dependent"). Suppose the random variable X has a normal distribution with mean 0 and variance 1, and set $Y = X^2$. Then Y is completely dependent on X: if you know the value of X you also know the value of Y. However,

$$C[X,Y] = C[X,X^2] = E[X^3] - E[X]E[X^2] = 0 - 0 \times 1 = 0,$$

since if X has a normal distribution with mean zero, then $E[X^3] = 0$. Thus X and Y are uncorrelated.   ▲

Example 2.3 ("Stockholm temperature"). The Stockholm temperature, recorded at the old Stockholm astronomical observatory, is one of the longest recorded temperature series that exists, starting 1756. After adjustment for the urban heat effect of the growing city and certain re-calibration one can calculate the average temperature for each month since January 1756; data are available from the web at the Bert Bolin Centre for Climate Research, [35]. The series contains 255 years of data, in all 3060 data points.

The series is not an example of a stationary process, since the mean value obviously depends on the season of the year. Let us for a moment forget about this non-stationarity and compute the covariance and correlation between mean temperatures taken τ months apart; see Section 2.5.3. Figure 2.3 shows the correlation function and a scatter plot of data nine months apart. The correlation is almost 0 for $\tau = 9$ despite a strong non-linear dependence.

A more sensible analysis of the temperature data would be to first subtract a seasonal average function from the temperature data, for example a cosine

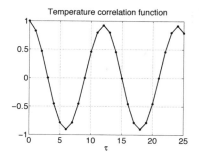
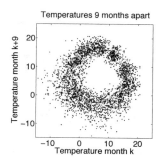

Figure 2.3 *Left: Correlation function of monthly mean temperature in Stockholm. Right: Scatter plot of mean temperatures taken nine months apart.*

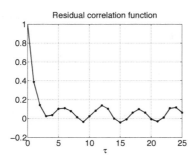
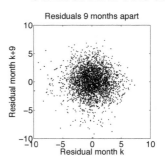

Figure 2.4 *Left: Correlation function for residuals of monthly mean temperature in Stockholm. Right: Scatter plot of residuals taken nine months apart.*

function. Without going into details, we show in Figure 2.4 the correlation function for the deviations from the mean curve, called the *residuals*. The scatter plot with time lag $\tau = 9$ shows no correlation, and the correlation function is almost zero there.   ▲

2.2.2 Simple properties and rules

The first and second order moments are linear and bi-linear, respectively. We formulate the following generalizations of the rules $E[aX + bY] = aE[X] + bE[Y]$ and $V[aX + bY] = a^2V(X) + b^2V[Y]$, which hold for uncorrelated random variables, X and Y.

> **Theorem 2.1.** *Let $a_1, \ldots, a_k$ and $b_1, \ldots, b_l$ be real constants, and let $X_1, \ldots, X_k$ and $Y_1, \ldots, Y_l$ be random variables in the same experiment, i.e., defined on a common sample space. Then*
>
> $$E\left[\sum_{i=1}^{k} a_i X_i\right] = \sum_{i=1}^{k} a_i E[X_i],$$
>
> $$V\left[\sum_{i=1}^{k} a_i X_i\right] = \sum_{i=1}^{k} \sum_{j=1}^{k} a_i a_j C[X_i, X_j],$$
>
> $$C\left[\sum_{i=1}^{k} a_i X_i, \sum_{j=1}^{l} b_j Y_j\right] = \sum_{i=1}^{k} \sum_{j=1}^{l} a_i b_j C[X_i, Y_j].$$

The rule for the covariance between sums of random variables, $C[\sum a_i X_i, \sum b_j Y_j]$, is easy to remember and use: the total covariance between two sums is a double sum of all covariances between pairs of one term from the first sum, and one term from the second sum.

We continue with some examples of covariance calculations.

Example 2.4. Assume X_1 and X_2 to be independent random variables and define a new variable $Z = X_1 - 2X_2$. The variance of Z is then

$$V[Z] = V[X_1 - 2X_2] = C[X_1 - 2X_2, X_1 - 2X_2]$$
$$= C[X_1, X_1] - 2C[X_1, X_2] - 2C[X_2, X_1] + 4C[X_2, X_2] = V[X_1] + 4V[X_2].$$

We also calculate the variance for the variable $Y = X_1 - 3$:

$$V[Y] = V[X_1 - 3] = C[X_1 - 3, X_1 - 3]$$
$$= C[X_1, X_1] - C[X_1, 3] - C[3, X_1] + C[3, 3] = V[X_1],$$

i.e., the same as for X_1. This is also intuitively clear; addition or subtraction of a constant shouldn't change the variance. ▲

Example 2.5. From a sequence $\{U_t\}$ of independent random variables with mean zero and variance σ^2, we construct a new process $\{X_t\}$ by

$$X_t = U_t + 0.5 \cdot U_{t-1}.$$

This is a "moving average" process, which is a topic in Chapter 7. By means of Theorem 2.1, we can calculate its mean value and covariance function. Of course, $m(t) = E[X_t] = E[U_t + 0.5 \cdot U_{t-1}] = 0$. For the covariance function, we have to work harder, and to keep computations under control, we do separate calculations according to the size of $t - s$. First, take $s = t$,

$$r(t,t) = V[X_t] = V[U_t + 0.5 \cdot U_{t-1}]$$
$$= V[U_t] + 0.5 \cdot C[U_{t-1}, U_t] + 0.5 \cdot C[U_t, U_{t-1}] + 0.5^2 \cdot V[U_{t-1}]$$
$$= \sigma^2 + 0 + 0 + 0.25\sigma^2 = 1.25\sigma^2,$$

where we used the fact that $V[U_t] = V[U_{t-1}] = \sigma^2$, and that U_t and U_{t-1} are independent, so $C[U_t, U_{t-1}] = C[U_{t-1}, U_t] = 0$. For $s = t + 1$ we get

$$r(t+1, t) = C[U_{t+1} + 0.5 \cdot U_t, U_t + 0.5 \cdot U_{t-1}]$$
$$= C[U_{t+1}, U_t] + 0.5 \cdot C[U_{t+1}, U_{t-1}] + 0.5 \cdot C[U_t, U_t] + 0.5^2 \cdot C[U_t, U_{t-1}]$$
$$= 0 + 0 + 0.5 \cdot V[U_t] + 0 = 0.5\sigma^2.$$

The case $s = t - 1$ gives the same result, and for $s \geq t + 2$ or $s \leq t - 2$, one easily finds that $r(s,t) = 0$. Process values $X(s)$ and $X(t)$ with time separation $|s-t|$ greater than one are therefore uncorrelated (they are even independent). All moving average processes share the common property that they have a finite correlation time: after some time lag the correlation is exactly zero. ▲

2.2.3   Interpretation of moments and moment functions

Expectation and the mean value function

The mean value function $m(t)$ of a stochastic process $\{X(t), t \in T\}$ is defined as the expected value of $X(t)$ as a function of time. From the law of large numbers in probability theory we know the precise meaning of this statement: for many independent repetitions of the experiment, i.e., many independent observations of the random $X(t)$, the arithmetic mean (i.e., the average) of the observations tends to be close to $m(t)$. But as its name suggests, one would also like to interpret it in another way: the mean function should say something about the average of the realization $x(t)$ over time. These two meanings of the word "average" are one of the subtle difficulties in the applications of stochastic process theory – later we shall try to throw some light upon the problem when we discuss how to estimate the mean value function, and introduce the concept of *ergodicity* in Section 2.5.2.

Correlation and covariance function

The second order moment functions, the covariance and correlation functions, measure the degree of linear dependence between process values at different times.

First, we discuss the correlation coefficient, $\rho = \rho[X,Y]$, between two random variables X and Y with positive variance,

$$\rho = \frac{C[X,Y]}{\sqrt{V[X]V[Y]}} = \frac{E[(X - m_X)(Y - m_Y)]}{\sqrt{V[X]V[Y]}}.$$

The correlation coefficient is a dimensionless constant, that remains unchanged after a change of scale: for constants $a > 0, c > 0, b, d$,

$$\rho(aX + b, cY + d) = \rho(X,Y).$$

It is easy to see that it is bounded to be between -1 and $+1$. To see this, we calculate the variance of $X - \lambda Y$ for the choice $\lambda = C[X,Y]/V[Y] = \rho\sqrt{V[X]/V[Y]}$. Since a variance is always non-negative, we have

$$0 \leq V[X - \lambda Y] = V[X] - 2\lambda C[X,Y] + \lambda^2 V[Y] = V[X](1 - \rho^2), \qquad (2.1)$$

which is possible only if $-1 \leq \rho \leq 1$.

Example 2.6 ("EEG"). The electroencephalogram is the graphic representation of spontaneous brain activity measured with electrodes attached to the scalp. Usually, EEG is measured from several channels at different positions

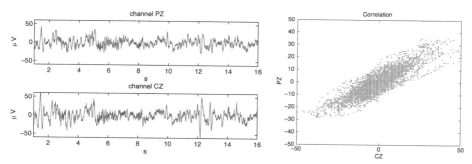

Figure 2.5 *Observed EEG-signals from two different channels (left), and scatter plot of x_t, y_t (right). The estimated correlation coefficient between X_t and Y_t is 0.9. (right)*

on the head. Channels at nearby positions will be strongly correlated, as can be seen in Figure 2.5, where the curves have a very similar appearance. But they are not identical, and using the samples as observations of two different stochastic processes, an estimate of the correlation coefficient between $X(t)$ and $Y(t)$ will be $\rho \approx 0.9$, i.e., rather close to one. Figure 2.5 also shows a scatter plot of the two signals, where the strong correlation is seen as the samples are distributed close to a straight line.   ▲

The covariance $E[(X - m_X)(Y - m_Y)]$ measures the degree of *linear* co-variation. If there is a tendency for observations of X and Y to be either both large or both small, compared to their expected values, then the product $(X - m_X)(Y - m_Y)$ is more often positive than negative, and the correlation is positive. If, on the other hand, large values of X often occur together with small values of Y, and vice versa, then the product is more often negative than positive, and the correlation is negative. From (2.1) we see that if the correlation coefficient is $+1$ or -1, there is an exact linear relation between X and Y, in the sense that there is a constant λ such that $V[X - \lambda Y] = 0$, which means that $X - \lambda Y$ is a constant, say a, i.e., $P(X = a + \lambda Y) = 1$. Repeated observations of the pair (X, Y) would then fall on a straight line. The closer the correlation is to ± 1 the closer to a straight line are the observations.

Figure 2.6 shows scatter plots of observations of two-dimensional normal variables with different degrees of correlation. As seen in the figure, there is quite a scatter around a straight line even with a correlation as high as 0.9.

Now back to the interpretation of the covariance function and its scaled version, the correlation function, of a stochastic process. If the correlation function, $\rho(s, t)$, attains a value close to 1 for some arguments s and t, then the realizations $x(s)$ and $x(t)$ will vary together. If on the other hand, the cor-

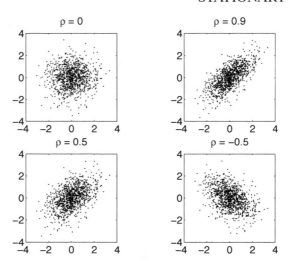

Figure 2.6 *Observations of two-dimensional normal variables X, Y with $E[X] = E[Y] = 0$ and $V[X] = V[Y] = 1$ for some different correlation coefficients ρ.*

relation is close to -1, the covariation is still strong, but goes in the opposite direction.

Example 2.7. This example illustrates how the sign of the correlation function is reflected in the variation of a stochastic process. Figure 2.7 shows realizations of a sequence of normal random variables $\{X_t\}$ with $m_t = 0$ and $r(t,t) = V[X_t] = 1$, and the covariance function (= correlation function since the variance is 1) $r(s,t) = \phi^{|s-t|}$, for four different ϕ-values. The realization with $\phi = 0.9$ shows rather strong correlation between neighboring observations, which becomes a little less obvious with $\phi = 0.5$. For $\phi = -0.5$ the correlation between observations next to each other, i.e., with $|s - t| = 1$, is negative, and this is reflected in the alternating signs in the realization.

Figure 2.8 illustrates the same thing in a different way. Pairs of successive observations (x_t, x_{t+k}), for $t = 1, 2, \ldots, n$, are plotted for three different time lags, $k = 1$, $k = 2$, $k = 5$. For $\phi = 0.9$ the correlation is always positive, but becomes weaker with increasing distance. For $\phi = -0.9$ the correlation is negative when the distance is odd, and positive when it is even, becoming weaker with increasing distance. ▲

2.3 Stationary processes

There is an unlimited number of ways to generate dependence in a stochastic process, and it is necessary to impose some restrictions and further assump-

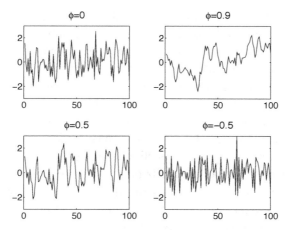

Figure 2.7 *Realizations of normal sequences with* $m(t) = 0$ *and* $r(s,t) = \phi^{|s-t|}$.

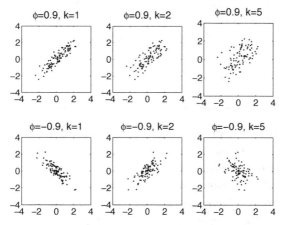

Figure 2.8 *Scatter plots of* (x_t, x_{t+k}) *for* $\phi = 0.9$ *and* $\phi = -0.9$.

tions in order to derive any useful and general properties about a process from observation of a time series. There are three main assumptions that are very much used to make the dependence manageable and which lead to scientifically meaningful models.

The first is the *Markov principle*, which says that the statistical distribution of what will happen between time s and time t depends on what happened up to time s only through the value at time s. This means that $X(s)$ is a *state variable* that summarizes the history before time s. In Chapters 6–8, we will meet some applications of this concept.

The second principle is a variation of the Markov principle, that assumes that the expected future change is 0, independently of how the process reached its present value. Processes with this property are called *martingales*, and they are central in *stochastic calculus* and financial statistics.

The third principle that makes the dependence manageable is the *stationarity principle*, and that is the topic of this book. In everyday language, the word *stationary* indicates something that does not change with time, or stays permanently in its position. In statistics, it means that certain *statistical properties* do not change. A random function is called "stationary" if the fluctuations have the same statistical distributions whenever one chooses to observe the process. The word stationary is mostly used for processes in time. For processes with a space parameter, for example a random surface or an image, the common term for this property is *homogeneous*.

For a process to be fully stationary, all statistical properties have to be unchanged with time. This is a very strict requirement, and to make life, and mathematics, simpler one can often be content with a weaker condition, namely that the mean value and covariances do not change.

2.3.1 Strictly stationary processes

Definition 2.3. *A stochastic process $\{X(t), t \in T\}$ is called strictly stationary if its statistical distributions remain unchanged after a shift of the time scale.*

Since the distributions of a stochastic process are defined by the finite-dimensional distribution functions,[1] we can formulate an alternative definition of strict stationarity:

If, for every n, every choice of times $t_1, \ldots, t_n \in T$ and every time lag τ such that $t_i + \tau \in T$, the n-dimensional random vector $(X(t_1 + \tau), \ldots, X(t_n + \tau))$ has the same multivariate distribution function as the vector $(X(t_1), \ldots, X(t_n))$, then the process $\{X(t), t \in T\}$ is said to be strictly stationary.

We now illustrate the stochastic time invariance property by an example where a shift in the time axis does not affect the distributions.

Example 2.8 ("Random phase and amplitude"). Let $A > 0$ and ϕ be independent random variables, and assume that the distribution of ϕ is uniform in the

---

[1]See Appendix C on Kolmogorov's existence theorem.

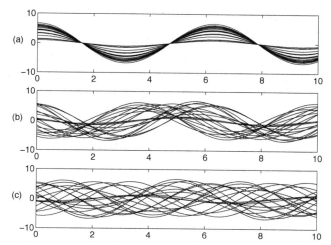

Figure 2.9 *Realizations of three different processes* $X(t) = A \cos(t + \phi)$ *with random amplitude A and with different phase distributions. (a): constant phase* $\phi = 0$. *(b): phase* ϕ *uniform between* 0 *and* π. *(c): uniform phase between* 0 *and* 2π. *Only the process in (c) is strictly stationary, where the choice of origin is irrelevant.*

interval $[0, 2\pi]$, i.e.,

$$f_{A,\phi}(x,y) = f_A(x)f_\phi(y) = f_A(x) \cdot \frac{1}{2\pi} \quad \text{for } x > 0, 0 \le y \le 2\pi.$$

The process $\{X(t), t \in \mathbb{R}\}$, defined as

$$X(t) = A \cos(t + \phi),$$

is strictly stationary; the distributions of $\{A \cos(t + \phi)\}$ are not affected by a shift of the time axis. Of course, the individual realizations are shifted, but since "random phase" means that one starts to observe the process "at a random time point," the distributions are unchanged. The top two plots in Figure 2.9 show realizations of $A \cos(t + \phi)$ when the phase is not completely random. ▲

If $\{X(t), t \in T\}$ is strictly stationary, then the marginal distribution of $X(t)$ is independent of t. Also the two-dimensional distributions of $(X(t_1), X(t_2))$ are independent of the absolute location of t_1 and t_2; only the distance $t_2 - t_1$ matters. As a consequence, the mean function $m(t)$ is constant, and the covariance function $r(s,t)$ is a function of $|t - s|$ only, not of the absolute location of s and t. Also higher order moments, like the third order moment, $E[X(s)X(t)X(u)]$, remain unchanged if one adds a constant time shift to s,t,u.

2.3.2   *Weakly stationary processes*

There are good reasons to study so called *weakly stationary* processes where the first and second order moments are time-invariant, i.e., the mean is constant and the covariances only depend on the time distance. The main reasons are, first of all, that weakly stationary Gaussian processes are automatically also strictly stationary, since their distributions are completely determined by mean values and covariances; see Section 1.3.2 and Chapter 5. Secondly, stochastic processes passed through linear filters are effectively handled by the first two moments of the input; see Chapters 6–8.

Definition 2.4. *If the mean function $m(t)$ is constant and the covariance function $r(s,t)$ is finite and depends only on the time difference $\tau = t - s$, the process $\{X(t), t \in T\}$ is called weakly stationary, or covariance stationary.*

Note that when we, in this book, say that a process is stationary, we mean that it is weakly stationary, if we do not explicitly say anything else.

Every strictly stationary process with finite variance is also weakly stationary. If it is Gaussian and weakly stationary, then it is also strictly stationary, but in general one can not draw such a conclusion. For a stationary process, we write m for the constant mean value and make the following simplified definition and notation for the covariance function.

Definition 2.5. *If $\{X(t), t \in T\}$ is a weakly stationary process with mean m, the covariance and correlation functions are defined as*

$$r(\tau) = \mathsf{C}[X(t), X(t+\tau)]$$
$$= \mathsf{E}[(X(t) - m)(X(t+\tau) - m)] = \mathsf{E}[X(t)X(t+\tau)] - m^2,$$
$$\rho(\tau) = \rho[X(t), X(t+\tau)] = r(\tau)/r(0).$$

In particular, the variance is $r(0) = \mathsf{V}[X(t)] = \mathsf{E}[(X(t) - m)^2]$.

We have used the same notation for the covariance function for a stationary as well as for a non-stationary process. No confusion should arise from

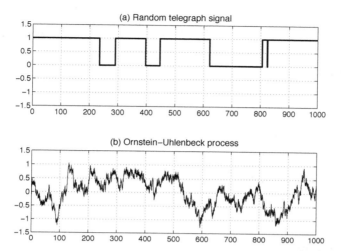

Figure 2.10 *Realizations of two processes with the same covariance function, $r(\tau) = \sigma^2 e^{-\alpha|\tau|}$: (a) random telegraph signal, (b) Ornstein-Uhlenbeck Gaussian process.*

this – one argument for the stationary case and two arguments for the general. For a stationary process, $r(s, s + \tau) = r(\tau)$.

The mean and covariance function can tell us much about how process values are connected, but they fail to provide detailed information about the sample functions, as is seen in the next example.

Example 2.9 ("Random telegraph signal" and "Ornstein-Uhlenbeck process"). Figure 2.10 shows realizations of two stationary processes with different distributions and quite different sample function behavior, but with exactly the same covariance function

$$r(\tau) = \sigma^2 e^{-\alpha|\tau|}.$$

The process in plot (a) is the "random telegraph signal," treated in Section 3.3, which jumps in a random fashion between the two levels 0 and 1, while the process in plot (b) is a Gaussian process with continuous, although rather irregular, realizations. It is called a "Gauss-Markov process" or an "Ornstein-Uhlenbeck process," and it will be studied more in Chapter 5. ▲

2.3.3 Important properties of the covariance function

All covariance functions share the following very important properties.

Theorem 2.2. *If $r(\tau)$ is the covariance function for a stationary process $\{X(t), t \in T\}$, then*

a. $r(0) = V[X(t)] \geq 0$,

b. $V[X(t+h) \pm X(t)] = E[(X(t+h) \pm X(t))^2] = 2(r(0) \pm r(h))$,

c. $r(0) = r(-\tau) = r(\tau)$,

d. $|r(\tau)| \leq r(0)$,

e. if $|r(\tau)| = r(0)$, for some $\tau \neq 0$, then r is periodic,

f. if $r(\tau)$ is continuous for $\tau = 0$, then $r(\tau)$ is continuous every-where.

Proof. (a) is clear by definition.

(b) Take the variance of the variables $X(t+h) + X(t)$ and $X(t+h) - X(t)$:

$$V[X(t+h) \pm X(t)] = V[X(t+h)] + V[X(t)] \pm 2C[X(t), X(t+h)]$$
$$= r(0) + r(0) \pm 2r(h) = 2(r(0) \pm r(h)).$$

(c) The covariance is symmetric in the arguments, so

$$r(-\tau) = C[X(t), X(t-\tau)] = C[X(t-\tau), X(t)] = r(t - (t-\tau)) = r(\tau).$$

(d) Since the variance of $X(t+h) \pm X(t)$ is non-negative regardless of the sign, part (b) gives that $r(0) \pm r(h) \geq 0$, and hence $|r(h)| \leq r(0)$.

(e) If $r(\tau) = r(0)$, part (b) gives that $V[X(t+\tau) - X(t)] = 0$, which, together with stationarity, implies that $X(t+\tau) = X(t)$ for all t, so $X(t)$ is periodic with period τ. If, on the other hand, $r(\tau) = -r(0)$, then $V[X(t+\tau) + X(t)] = 0$, and $X(t+\tau) = -X(t)$ for all t, and $X(t)$ is periodic with period 2τ. Finally, it is easy to see that if $X(t)$ is periodic, then also $r(t)$ is periodic.

(f) We consider the increment of the covariance function at t,

$$(r(t+h) - r(t))^2 = (C[X(0), X(t+h)] - C[X(0), X(t)])^2$$
$$= (C[X(0), X(t+h) - X(t)])^2, \qquad (2.2)$$

where we used that $C[U,V] - C[U,W] = C[U, V-W]$. Further, since correlation is absolutely bounded by one,

$$(C[Y,Z])^2 \leq V[Y]V[Z].$$

Applied to the right hand side in (2.2) this yields, according to (b),

$$(r(t+h) - r(t))^2 \leq \mathsf{V}[X(0)] \cdot \mathsf{V}[X(t+h) - X(t)] = 2r(0)(r(0) - r(h)).$$

If $r(\tau)$ is continuous for $\tau = 0$, then, in the right hand side, $r(0) - r(h) \to 0$ as $h \to 0$. Then also the left hand side $(r(t+h) - r(t)) \to 0$, and hence $r(\tau)$ is continuous at $\tau = t$. ∎

The theorem is important since it restricts the class of functions that can be used as covariance functions for real stationary processes. For example, a covariance function must be symmetric and attain its maximum at $\tau = 0$. It must also be continuous if it is continuous at the origin, which excludes for example the function in Figure 2.11. In Chapter 4, Theorem 4.2, we shall give a definite answer to the question of what functions can appear as covariance functions.

2.4 Random phase and amplitude

We spend this entire section on processes that are built by simple harmonic functions, where only the amplitudes and phases are random. We first deal with the random harmonic oscillator.

2.4.1 A random harmonic oscillation

Let $A > 0$ and ϕ be independent random variables, with ϕ uniform over $[0, 2\pi]$. Also assume A to have finite variance, and write $\mathsf{E}[A^2] = 2\sigma^2$. Fix $f_0 > 0$ and, as in Example 1.12, define the stochastic process $\{X(t), t \in T\}$ by

$$X(t) = A\cos(2\pi f_0 t + \phi). \tag{2.3}$$

It will be a stationary process with realizations that are periodic functions with period $1/f_0$ and *frequency* f_0. The unit of f_0 is [(time unit)$^{-1}$]. One can

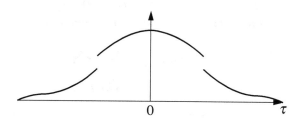

Figure 2.11 *A function discontinuous at $\tau \neq 0$ cannot be a covariance function.*

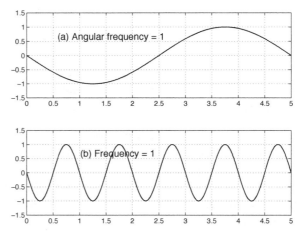

Figure 2.12 *(a): Angular frequency* $\omega_0 = 1$ *(frequency* $f_0 = 1/2\pi$*). (b): (Angular frequency* $\omega_0 = 2\pi$*), frequency* $f_0 = 1$.

also use *angular frequency*, defined as $\omega_0 = 2\pi f_0$, which has units of [rad (time unit)$^{-1}$], and define the process as

$$X(t) = A\cos(\omega_0 t + \phi). \qquad (2.4)$$

When the time unit is second (s), the frequency f_0 has the unit of *Hertz*, Hz = [s^{-1}], while the angular frequency ω_0 has the unit [rad s^{-1}].

Figure 2.12 illustrates the two functions $t \mapsto \cos(\omega_0 t + \phi)$ for $\omega_0 = 1$ and $t \mapsto \cos(2\pi f_0 t + \phi)$ for $f_0 = 1$, which appear in (2.3) and (2.4). The figure shows the case $\phi = 0$. In this book, we choose the representation (2.3) with frequency, not angular frequency, to allow for easier physical interpretation.

Theorem 2.3. *The random harmonic function* $X(t) = A\cos(2\pi f_0 t + \phi)$ *is a stationary process if* A *and* ϕ *are independent and* ϕ *is uniformly distributed in* $[0, 2\pi]$. *It has mean zero,* $E[X(t)] = 0$, *and variance and covariance function,*

$$V[X(t)] = \frac{1}{2}E[A^2] = \sigma^2,$$
$$r(\tau) = \sigma^2 \cos 2\pi f_0 \tau. \qquad (2.5)$$

Proof. The random phase ϕ has the effect that the value $X(t)$ is equal to what we get if we sample a cosine function at "a completely random point of time." Since, taken over a full period, the "average" of a cosine function is 0, it is hence intuitively clear that $E[X(t)]$ should be equal to 0. Thus, the mean function is constant: $m = m(t) = E[X(t)] = 0$. Set in formulas, since A and ϕ are assumed to be independent,

$$E[X(t)] = E[A\cos(2\pi f_0 t + \phi)] = E[A] \cdot E[\cos(2\pi f_0 t + \phi)].$$

Here the last factor equals 0:

$$E[\cos(2\pi f_0 t + \phi)] = \int \cos(2\pi f_0 t + y) f_\phi(y)\,dy$$

$$= \int_0^{2\pi} \cos(2\pi f_0 t + y) \frac{1}{2\pi}\,dy = \left[\frac{1}{2\pi}\sin(2\pi f_0 t + y)\right]_{y=0}^{2\pi}$$

$$= \frac{1}{2\pi}\{\sin(2\pi f_0 t + 2\pi) - \sin(2\pi f_0 t)\} = 0.$$

For the covariance function, we find

$$E[X(s)X(t)] = E[A^2 \cos(2\pi f_0 s + \phi)\,\cos(2\pi f_0 t + \phi)]$$

$$= E[A^2] \cdot E[\cos(2\pi f_0 s + \phi)\cos(2\pi f_0 t + \phi)].$$

Here, the second factor is $\frac{1}{2}\cos(2\pi f_0(s - t))$:

$$E[\cos(2\pi f_0 s + \phi)\cos(2\pi f_0 t + \phi)]$$

$$= \int \cos(2\pi f_0 s + y)\cos(2\pi f_0 t + y) f_\phi(y)\,dy$$

$$= \int_0^{2\pi} \cos(2\pi f_0 s + y)\cos(2\pi f_0 t + y) \frac{1}{2\pi}\,dy.$$

Using the formula $\cos\alpha \cdot \cos\beta = \frac{1}{2}(\cos(\alpha+\beta) + \cos(\alpha - \beta))$, we can reduce the expression to

$$\frac{1}{2\pi}\int_0^{2\pi} \frac{1}{2}(\cos(2\pi f_0(s+t) + 2y) + \cos(2\pi f_0(s - t)))\,dy = \frac{1}{2}\cos(2\pi f_0(s - t)).$$

We have found that

$$E[X(s)X(t)] = \frac{1}{2}E[A^2]\cos(2\pi f_0(s - t)),$$

so the process has covariance function

$$r(\tau) = \frac{1}{2}E[A^2]\cos 2\pi f_0 \tau = \sigma^2 \cos 2\pi f_0 \tau,$$

and since the mean function equals 0 it is hence stationary.

The variance is

$$V[X(t)] = r(0) = \frac{1}{2}E[A^2] = \sigma^2,$$

which explains the notation $2\sigma^2$ for $E[A^2]$.

Actually, there is a much simpler way to find the variance. Since $E[X(t)] = 0$, the variance is $E[X(t)^2] - 0^2 = E[A^2] \cdot E[(\cos(2\pi f_0 t + \phi))^2] = E[A^2] \cdot E[(\sin(2\pi f_0 t + \phi))^2]$, because a randomly shifted cosine function can not be distinguished from a randomly shifted sine function. But $\cos^2 x + \sin^2 x = 1$, so the two expectations are equal to one half each.   ∎

2.4.2   Superposition of random harmonic oscillations

A cosine function with random amplitude and phase is not very useful as a model for a random function of time. However, adding many independent such cosine functions will make a big difference, and lead to very general and useful stochastic processes.

Choose n different (non-random) frequencies $f_1, \ldots, f_n$, and let $A_1, \ldots, A_n$ and $\phi_1, \ldots, \phi_n$ be mutually independent random variables. Let all ϕ_k be uniform in the interval $[0, 2\pi]$, and write $E[A_k^2] = 2\sigma_k^2$. Then, the sum $X(t) = \sum_{k=1}^{n} A_k \cos(2\pi f_k t + \phi_k)$ defines a stationary process with mean value function 0. Since the separate terms in $X(t)$ are independent they are also uncorrelated, and Theorem 2.1 gives that the covariance function is $r(\tau) = \sum_{k=1}^{n} \sigma_k^2 \cos 2\pi f_k \tau$.

In some applications, there are reasons to include also a random constant A_0 with variance $V[A_0] = E[A_0^2] = \sigma_0^2$, independent of all other A_k and ϕ_k. The total process,

$$X(t) = A_0 + \sum_{k=1}^{n} A_k \cos(2\pi f_k t + \phi_k), \tag{2.6}$$

then consists of n independent cosine function with random amplitude and phase (constant in time), added to a randomly chosen average, which stays constant over the whole realization. The total covariance function is

$$r(\tau) = \sigma_0^2 + \sum_{k=1}^{n} \sigma_k^2 \cos 2\pi f_k \tau. \tag{2.7}$$

The superposition of random cosine function can be extended to a sum of infinitely many terms, but for that we have to introduce a notion of stochastic convergence. This will be dealt with in Section 6.3.

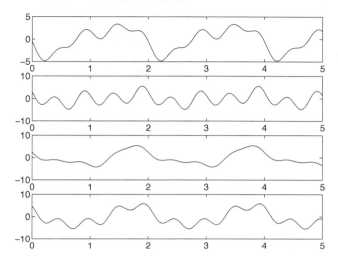

Figure 2.13 *Four "random" functions generated by the same random mechanism:* $X(t) = \sum_1^3 A_k \cos(2\pi f_k t + \phi_k)$, $E(A_1^2) = E(A_3^2) = 2$, $E(A_2^2) = 4$, $f_1 = 0.5, f_2 = 1, f_3 = 2$.

Example 2.10 ("Three cosine functions"). Already a sum of three cosine functions, with different frequencies, amplitudes, and phases, can look quite "random" at first glance. Figure 2.13 shows four realizations of the same process. ▲

2.4.3   Power and average power

In many applications of stationary processes, the variance $V[X(t)] = r(0)$ or sometimes $E[X(t)^2]$ are quantities of main interest. For example, if $X(t)$ represents the potential difference ("voltage") over a resistance R, the electric power in the circuit is $P = X(t)^2/R$, and on the average it is proportional to $E[X(t)^2]$. As another example, we can take $X(t)$ as the varying height of the sea surface around the mean sea level. Then, $E[X(t)^2]$ is proportional to the amount of energy per surface area available as wave power.

The model (2.6) is particularly useful to describe a fluctuating electric potential over a resistance. Then, A_0 is a DC ("Direct current") component and the other terms are AC ("Alternating current") components with different frequencies. We assume that all amplitudes, including A_0, are chosen at random and that $E[A_0] = 0$, so $E[X(t)] = 0$. If $X(t)$ is the potential difference over a resistance of $R = 1\,\Omega$, $r(0) = E[X(t)^2]$ is the expected power that the current develops in the resistance.

We will now see how $r(0) = \sigma_0^2 + \sum_{k=1}^n \sigma_k^2$ relates to the average power that is actually developed in the circuit when the potential varies according to a realization of (2.6). Consider the time averaged power $E_T = \frac{1}{T}\int_0^T X(t)^2\, dt$ over the interval $[0,T]$. By direct application of trigonometric identities we get

$$E_T = \frac{1}{T}\int_0^T X(t)^2\, dt = \frac{1}{T}\left\{ TA_0^2 + 2\sum_{k=1}^n A_0 A_k \int_0^T \cos(2\pi f_k t + \phi_k)\, dt \right.$$

$$+ \sum_{k=1}^n A_k^2 \int_0^T \cos^2(2\pi f_k t + \phi_k)\, dt$$

$$\left. + 2\sum_{l=1}^{n-1}\sum_{k=l+1}^n A_k A_l \int_0^T \cos(2\pi f_l t + \phi_l)\cos(2\pi f_k t + \phi_k)\, dt \right\}.$$

It is easy to see that $T^{-1}\int_0^T \cos(2\pi f_k t + \phi_k)\, dt \to 0$ as $T \to \infty$, and also, for $f_k \neq f_l$,

$$T^{-1}\int_0^T \cos(2\pi f_l t + \phi_l)\cos(2\pi f_k t + \phi_k)\, dt \to 0,$$

$$T^{-1}\int_0^T \cos^2(2\pi f_k t + \phi_k)\, dt \to 1/2.$$

Thus, as $T \to \infty$,

$$E_T \to A_0^2 + \sum_{k=1}^n A_k^2/2, \tag{2.8}$$

i.e., the average power over time in a realization of the process is $A_0^2 + \sum_{k=1}^n A_k^2/2$. We now explain how this relates to $r(0)$.

Since $E[A_0^2] = \sigma_0^2$ and $E[A_k^2/2] = \sigma_k^2$ for $k = 1,\ldots,n$, we have, from (2.7) and (2.8),

$$r(0) = \sigma_0^2 + \sum_{k=1}^n \sigma_k^2 = E\left[A_0^2 + \frac{1}{2}\sum_{k=1}^n A_k^2 \right] = E\left[\lim_{T\to\infty} E_T \right], \tag{2.9}$$

i.e., the expected value of the long term average power is equal to $r(0)$. Be aware that both the average power $(1/T)\int_0^T X(t)^2\, dt$ over $[0,T]$ and its long term limit as $T \to \infty$ are random variables, so it is only meaningful to talk about the expectation of the average power.

The equality (2.9) is important. It says that $r(0)$ is equal to both the expected momentary power (i.e., the variance of $X(t)$) and the expectation of the long term average power in a single realization:

$$r(0) = E[X(t)^2] = E\left[\lim_{T\to\infty} \frac{1}{T}\int_0^T X(t)^2\, dt \right] = \lim_{T\to\infty} E[E_T].$$

This is true for all stationary processes with $E[X(t)] = 0$.

In Section 2.5.4, we shall introduce a more general property, *ergodicity*, which implies that, for *every realization* $x(t)$,

$$r(0) = \lim_{T \to \infty} E_T = \lim_{T \to \infty} \frac{1}{T} \int_0^T x(t)^2 \, dt.$$

This property is often described as: "For an ergodic process, the ensemble average (= expectation) of the power is equal to the time average." In the future, we will often call $r(0)$ "the average power of the process." Here, we assumed that the process mean was zero, and we will do so in the future, when $r(0)$ is referred to as the average power.

2.5 Estimation of mean value and covariance function

So far, we have described the most simple characteristics of a stationary process, the mean value and covariance function, and these were defined in a probabilistic way, as expectations that can be *calculated* from a distribution or probability density function. They were used to *describe* and *predict* what can be expected from a realization of the process. We did not pay much attention to how the model should be chosen.

However, as always in statistics, model selection, *parameter estimation*, and *model fitting* are needed in order to find a good model that can be used in practice for prediction of a time series and for simulation and theoretical analysis. In this section, we discuss the properties of the natural estimators of the mean value and covariance function of a stationary process $\{X_n, n = 1, 2, \ldots\}$.

Before we start with the technical details, just a few words about how to think about parameter estimation, model fitting, and data analysis.

Formal parameter estimation: On the formal statistical level, a model is postulated beforehand, and the data are assumed to be observations of random variables for which distributions are known, apart from some unknown parameters. The task is to suggest a function of the observations that has good properties as an estimate of the unknown parameters. The model is not questioned, and the procedure is evaluated according to how close the estimates are to the true parameter values.

Model selection: On this level, one seeks to find which model, among many possible alternatives, that is most likely to have produced the data. This includes parameter estimation as well as model choice. The procedure is evaluated from its ability to reproduce the observations and to predict future observations.

Model fitting: This is the most realistic situation in practice. No model is
assumed to be "true," and the task is to find a model and parameters that
best describes observed data, and that well predicts the future.

In this course, we will stay with the first level, and study the properties of estimation procedures, but the reader is reminded that this is a formal approach.
No model should be regarded as true, unless there are some external reasons
to assume a specific structure, for example derived from physical principles,
and perhaps not even then! In this section we deal only with estimation of
the mean value and the covariance function. Parameter estimation in special
models is the topic of Chapter 7; see further the time series analysis literature,
e.g., [10, 11].

2.5.1   Estimation of the mean value function

Let $\{X_n, n = 1, 2, \ldots\}$ be a weakly stationary sequence. It need not be strictly
stationary, and we make no specific assumption about its distributions, apart
from the assumption that it has a constant, but unknown, mean value m and a
(known or unknown) covariance function $r(\tau)$. Let $x_1, \ldots, x_n$ be observations
of $X_1, \ldots, X_n$.[2]

Theorem 2.4. (a) The arithmetic mean $\widehat{m}_n = \frac{1}{n} \sum_{t=1}^{n} X_t$ is an unbiased estimator of the mean value m of the process, i.e., $\mathsf{E}[\widehat{m}_n] = m$,
regardless of the distribution.

(b) If the infinite series $\sum_{\tau=0}^{\infty} r(\tau)$ is convergent, then the asymptotic
variance of $\widehat{m}_n$ is given by

$$\lim_{n \to \infty} n\mathsf{V}[\widehat{m}_n] = \sum_{\tau=-\infty}^{\infty} r(\tau) = r(0) + 2\sum_{1}^{\infty} r(\tau), \qquad (2.10)$$

which means that, for large n, the variance of the mean value estimator is $\mathsf{V}[\widehat{m}_n] \approx \frac{1}{n} \sum_{\tau} r(\tau)$.

(c) Under the condition in (b), $\widehat{m}_n$ is a consistent estimator of m as
$n \to \infty$ in the sense that $\mathsf{E}[(\widehat{m}_n - m)^2] \to 0$, and also $\mathsf{P}(|\widehat{m}_n - m| >
\varepsilon) \to 0$, for all $\varepsilon > 0$.

---

[2]As customary in statistics, we often do not make any notational distinction between an
estimator, as function of random variables, and its value for a sample of the variables. For
example, $\widehat{m}_n = \frac{1}{n} \sum_{1}^{n} X_t$ is an estimator and $\widehat{m}_n = \frac{1}{n} \sum_{1}^{n} x_t$ is its value, computed from observations.

Proof. (a) The expectation of a sum of random variables is equal to the sum of the expectations; this is true regardless of whether the variables are dependent or not. Therefore, $E[\widehat{m}_n] = \frac{1}{n}\sum_{t=1}^{n} E[X_t] = m$.

(b) We calculate the variance by means of Theorem 2.1. From the theorem,

$$V[\widehat{m}_n] = \frac{1}{n^2} V\left[\sum_{t=1}^{n} X_t\right] = \frac{1}{n^2}\sum_{s=1}^{n}\sum_{t=1}^{n} C[X_s, X_t] = \frac{1}{n^2}\sum_{s=1}^{n}\sum_{t=1}^{n} r(s-t).$$

Now, if we sum along the diagonals and collect all the $n - |u|$ terms with $s - t = u$ we get $V[\widehat{m}_n] = \frac{1}{n^2}\sum_{u=-n+1}^{n-1}(n - |u|)r(u)$. Hence, using that $r(u) = r(-u)$ we get that

$$nV[\widehat{m}_n] = r(0) + \frac{2}{n}\sum_{u=1}^{n-1}(n - u)r(u). \qquad (2.11)$$

Now, if $\sum_{t=0}^{\infty} r(t)$ is convergent, $S_n = \sum_{t=0}^{n-1} r(t) \to \sum_{t=0}^{\infty} r(t) = S$, say, which implies that $\frac{1}{n}\sum_{k=1}^{n} S_k \to S$, as $n \to \infty$.[3] Thus,

$$\frac{1}{n}\sum_{u=0}^{n-1}(n - u)r(u) = \frac{1}{n}\sum_{k=1}^{n} S_k \to S,$$

and this, together with (2.11), gives the result, i.e., $nV[\widehat{m}_n)] \to r(0) + 2\sum_{1}^{\infty} r(u) = \sum_{-\infty}^{\infty} r(u)$.

(c) The first statement follows from (a) and (b), since $E[(\widehat{m}_n - m)^2] = V[\widehat{m}_n] + E[\widehat{m}_n - m]^2 \to 0$. The second statement is a direct consequence of Chebyshev's inequality,

$$P(|\widehat{m}_n - m| > \varepsilon) \leq \frac{E[(\widehat{m}_n - m)^2]}{\varepsilon^2}. \qquad \blacksquare$$

The consequences of the theorem are extremely important for data analysis with dependent observations. A positive correlation between successive observations tends to increase the variance of the mean value estimate, and make it more uncertain. We will give two examples on this, first to demonstrate the difficulties with visual interpretation, and then use the theorem for a more precise analysis.

Example 2.11 ("Dependence can deceive the eye"). Figure 2.14 shows 40 realizations of 25 successive data points in two stationary processes with discrete time. In the upper diagram, the data are dependent and each successive

---

[3]If $x_n \to x$ then also $(1/n)\sum_{k=1}^{n} x_k \to x$.

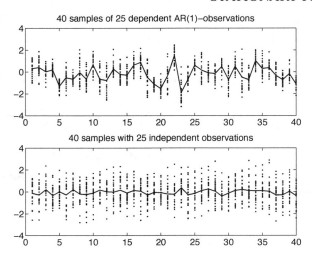

Figure 2.14 *Upper diagram: 40 samples of 25 dependent AR(1)-variables, $X_{t+1} = 0.9X_t + e_t$. Lower diagram: independent data Y_t with $V[Y_t] = V[X_t] = 1$. Also, $E[Y_t] = E[X_t] = 0$. Solid lines connect the averages of the 25 data points in each series. Note that the data in each of the 40 samples are represented by 25 dots in the 40 vertical columns in the upper and lower diagrams.*

data value contains part of the previous value, $X_{t+1} = 0.9X_t + e_t$, where the e_t are independent normal variables. (This is an AR(1)-process, an autoregressive process, which we will study in detail in Chapter 7.) In the lower diagram, all variables are independent normal variables $Y_t = e_t'$. Variances are chosen so that $V[X_t] = V[Y_t]$. The solid thick line connects the average of the 25 points in each sample.

One should note that the observations in the AR(1)-process within each sample are less spread out than those in the samples with independent data. However, even if there is less variation in each sample, the observed average values are more variable for the AR(1)-process. The calculated standard deviation in the 40 dependent samples is 0.68 on the average, while the average standard deviation is 0.99 for independent data. Therefore, if one had only one sample to analyze, and didn't notice the dependence, one could be led to believe that a dependent sample would give better precision in the estimate of the overall mean level. But it is just the opposite in this example; the more spread out sample gives better precision.

The reader is encouraged to repeat the experiment with the model

$$X_{t+1} = -0.9X_t + e_t. \qquad \blacktriangle$$

Example 2.12 ("How many data samples are necessary?"). How many data points should be sampled from a stationary process in order to obtain a specified precision in an estimate of the mean value function? As we saw from the previous example, the answer depends on the covariance structure. Successive time series data taken from nature often exhibit a very simple type of dependence, and often the covariance function can be approximated by a geometrically decreasing function, $r(\tau) = \sigma^2 e^{-\alpha|\tau|}$.

Suppose that we want to estimate the mean value m of a stationary sequence $\{X_t\}$, and that we have reasons to believe that the covariance function is of the type above with $\alpha = 1$, so that $r(\tau) = \sigma^2 e^{-|\tau|}$. We estimate m by the average $\widehat{m}_n = \bar{x} = \frac{1}{n}\sum_1^n x_k$ of observations of $X_1, \ldots, X_n$.

If the variables had been uncorrelated, the standard deviation of $\widehat{m}_n$ would have been $\sigma/\sqrt{n}$. We use Theorem 2.4, and calculate

$$\sum_{t=-\infty}^{\infty} r(t) = \sigma^2 \sum_{t=-\infty}^{\infty} e^{-|t|} = \sigma^2\left(1 + 2\sum_{t=1}^{\infty} e^{-|t|}\right)$$

$$= \sigma^2\left(1 + 2\frac{1/e}{1 - 1/e}\right) = \sigma^2 \frac{e+1}{e-1} < \infty,$$

so $\widehat{m}_n$ is unbiased and consistent. If n is large, the variance is

$$V[\widehat{m}_n] \approx \frac{\sigma^2}{n} \cdot \frac{e+1}{e-1}$$

and the standard deviation

$$D[\widehat{m}_n] \approx \frac{\sigma}{\sqrt{n}}\left(\frac{e+1}{e-1}\right)^{1/2} = \sigma\frac{1.471}{\sqrt{n}}.$$

We see that in this example the positively correlated data give almost 50% larger standard deviation in the m-estimate than uncorrelated data. To compensate this reduction in precision, it is necessary to measure over a longer time period; more precisely, to obtain the same variance as in an independent sample of length n one has to obtain $1.471^2 n \approx 2.16n$ measurements from the dependent sequence.

The constant α determines the decay of correlation. For a general exponential covariance function, and setting $\theta = e^{-\alpha}$, we have that $r(\tau) = \sigma^2 e^{-\alpha|\tau|} = \sigma^2 \theta^{|\tau|}$, and the asymptotic standard deviation is, for large n,

$$D[\widehat{m}_n] \approx \frac{\sigma}{\sqrt{n}}\left(\frac{e^\alpha + 1}{e^\alpha - 1}\right)^{1/2} = \frac{\sigma}{\sqrt{n}}\left(\frac{1+\theta}{1-\theta}\right)^{1/2}, \qquad (2.12)$$

which can be quite large for θ near 1. As a rule of thumb, if one wants to get the same variance of the mean in a sample from a stochastic process with this covariance function as in a sample of independent random variables with the same variance, then one has to increase the number of observations by a factor $(K^{1/\tau_K}+1)/(K^{1/\tau_K}-1)$, where τ_K is the time lag where the correlation has decreased from one to $1/K$.   ▲

Example 2.13 ("Oscillating data can decrease variance"). If the decay parameter θ in $r(\tau) = \sigma^2\theta^{|\tau|}$ is negative, the observations oscillate around the mean value, and the variance of the observed average will be smaller than for independent data. The "errors" tend to compensate each other. With $\theta = -1/e$, instead of $\theta = 1/e$ as in the previous example, the standard deviation of the observed mean is $\mathsf{D}[\widehat{m}_n] \approx \frac{\sigma}{\sqrt{n}}\left(\frac{e-1}{e+1}\right)^{1/2} = \sigma\frac{0.6798}{\sqrt{n}}$.   ▲

If dependence is "not too strong" (the conditions needed to ensure this are more restrictive than the conditions of Theorem 2.4, see [11, Sec. 6.4]), the arithmetic mean of a stationary sequence approximately has a normal distribution. In the next example we first assume that the process is Gaussian, and then use this to conclude that the results are valid also for more general processes than Gaussian ones.

Example 2.14 ("Confidence interval for the mean"). If the process $\{X_t\}$ in Example 2.12 is a Gaussian process, the estimator $\widehat{m}_n = \frac{1}{n}\sum_1^n X_t$ has a normal distribution with expectation m and approximate standard deviation $\mathsf{D}[\widehat{m}_n] \approx \frac{\sigma}{\sqrt{n}}1.471$, i.e., $\widehat{m}_n \sim N(m, \mathsf{D}[\widehat{m}_n]^2)$. This means, for example, that

$$P(m - q_{\alpha/2}\mathsf{D}[\widehat{m}_n] \leq \widehat{m}_n \leq m + q_{\alpha/2}\mathsf{D}[\widehat{m}_n]) = 1 - \alpha, \qquad (2.13)$$

where $q_{\alpha/2}$ is a quantile in the standard normal distribution:

$$P(-q_{\alpha/2} \leq Y \leq q_{\alpha/2}) = 1 - \alpha,$$

if $Y \sim N(0,1)$. Rearranging the inequality in (2.13) we get an equivalent event with

$$P(\widehat{m}_n - q_{\alpha/2}\mathsf{D}[\widehat{m}_n] \leq m \leq \widehat{m}_n + q_{\alpha/2}\mathsf{D}[\widehat{m}_n]) = 1 - \alpha.$$

We obtain a confidence interval for m,

$$I_m: \quad \{\widehat{m}_n - q_{\alpha/2}\mathsf{D}[\widehat{m}_n], \widehat{m}_n + q_{\alpha/2}\mathsf{D}[\widehat{m}_n]\}, \qquad (2.14)$$

with confidence level $1 - \alpha$. The interpretation is this: If the experiment "observe $X_1, \ldots, X_n$ and calculate $\widehat{m}_n$ and the interval I_m according to (2.14)" is repeated many times, some of the so constructed intervals will be "correct" and cover the true m-value, and others will not. In the long run, the proportion of correct intervals will be $1 - \alpha$.

Suppose now that we have observed the first 100 values of X_t, and got the sum $x_1 + \cdots + x_{100} = 34.9$. An estimate of m is $\widehat{m}_n = 34.9/100 = 0.349$ and a 95% confidence interval for m is, if $\sigma = 1$,

$$0.349 \pm q_{0.025} \cdot \frac{1.471}{\sqrt{100}} = 0.349 \pm 1.96 \cdot 0.1471 = 0.349 \pm 0.288.$$

(For a confidence level of 0.95 we must set $\alpha = 0.05$ and use $\lambda_{\alpha/2} = \lambda_{0.025} = 1.96$.) Thus, we found the 95% confidence interval for m to be $(0.061, 0.637)$.

Compare this with the interval constructed under the assumption of independent observations:

$$0.349 \pm q_{0.025} \cdot \frac{1}{\sqrt{100}} = 0.349 \pm 1.96 \cdot 0.1 = 0.349 \pm 0.2,$$

see Figure 2.14, which shows the increased uncertainty in the estimate for positively dependent variables.

The analysis was based on the normality assumption, but it is approximately valid also for moderately non-normal data. Since $\widehat{m}_n$ is the average of random variables one can apply a version of the *central limit theorem* for a strictly stationary process in which the variables are not too strongly dependent. An example of such dependence is the notion of *m-dependence*, which means that variables, more than m time units apart, are independent. For precise conditions, see [11, Sec. 6.4]. ▲

2.5.2   Ergodicity

Ensemble average

The expectation $m = \mathsf{E}[X]$ of a random variable X is sometimes also called the *ensemble average* of the experiment, and it is obtained as the average of the possible outcomes of the experiment, weighted by their probabilities.

If one makes many independent repetitions of the experiment, so that one gets independent observations $x_1, x_2, \ldots$ of X, then by the law of large numbers the average of these observations will converge to m:

$$\widehat{m}_n = \frac{1}{n} \sum_{k=1}^{n} x_k \to m, \text{ when } n \to \infty. \tag{2.15}$$

In other words, the average of the results from many independent repetitions of the experiment converges to the ensemble average.

Time average

Suppose now that instead an entire time series $\{X_t\}$ is the result of the experiment. If one makes many independent repetitions of the experiment, so that one obtains many independent observations of the time series, then, exactly as above, by the law of large numbers, the average of the first observations in these time series will converge to $m(1)$, the average of the second observations in these time series will converge to $m(2)$, and so on, where $m(t) = E[X_t]$.

In general, in a stochastic process the expectation $m(t)$ is different for different time points t. However, if the process is stationary, strictly or weakly, all the variables have the same distribution and hence also the same expectation. Call this common expectation m, so that $m = E[X_1] = E[X_2] = \ldots$, etc.

The following then is a natural question: Can one replace repeated independent observations of X_1 (or X_2 or ...) in the estimation of $m = E[X_1]$ (or $m = E[X_2]$ or ...), with observation of just *one single realization* $x_1, x_2, \ldots$ of the process, and estimate the common expectation m by the *time average* $\widehat{m}_n = \frac{1}{n}\sum_{t=1}^{n} x_t$? Thus, in the terminology above, does the time average converge to the ensemble average m? The answer is yes, for some processes. These processes are called *ergodic*, or more precisely *linearly ergodic*.

A linearly ergodic stationary sequence is a stationary process $\{X_n, n = 1, 2, \ldots\}$ where the common mean value m (= the ensemble average) can be consistently estimated by the time average,

$$\frac{x_1 + \cdots + x_n}{n} \to m, \text{ as } n \to \infty,$$

when $x_1, \ldots, x_n$ are observations in one single realization of $\{X_t\}$.

From Theorem 2.4(c) we get a sufficient condition for a stationary sequence to be linearly ergodic: $\sum_0^\infty r(\tau) < \infty$ implies that $P(|\widehat{m}_n - m| > \varepsilon) \to 0$ for all $\varepsilon > 0$.

If the process is linearly ergodic, one can estimate the expectation of any linear function $aX_t + b$ of X_t by the corresponding time average $\frac{1}{n}\sum_{t=1}^{n}(ax_t + b)$. This explains the name, "linearly" ergodic.

The essence of an ergodic process is that everything that can conceivably happen in repeated experiments also happens already in one single realization, if it is extended indefinitely, and that it happens in the proportions given by the probabilities for the different outcomes of the experiment.

2.5.3   Estimating the covariance function

The covariance function $r(\tau) = E[(X_t - m)(X_{t+\tau} - m)] = E[X_t X_{t+\tau}] - m^2$ of a stationary sequence $\{X_n\}$ is estimated by the appropriate sample covariance. First assume m is known.

Theorem 2.5. *The estimator*

$$\widehat{r}_n(\tau) = \frac{1}{n} \sum_{t=1}^{n-\tau} (x_t - m)(x_{t+\tau} - m), \qquad \tau \geq 0, \qquad (2.16)$$

is asymptotically unbiased, i.e., $E[\widehat{r}_n(\tau)] \to r(\tau)$ *when* $n \to \infty$.

Proof. The proof is by direct calculation, $E[\widehat{r}_n(\tau)] = \frac{1}{n} \sum_{t=1}^{n-\tau} E[(X_t - m)(X_{t+\tau} - m)] = \frac{1}{n} \sum_{t=1}^{n-\tau} r(\tau) = \frac{1}{n}(n-\tau)r(\tau) \to r(\tau)$ when $n \to \infty$.   ∎

Remark 2.2. *Since* $\widehat{r}_n(\tau)$ *is asymptotically unbiased it is consistent as soon as its variance goes to 0 when* $n \to \infty$. *Theorem 2.4 applied to the process* $Y_t = (X_t - m)(X_{t+\tau} - m)$ *will yield the result, since* Y_t *has expectation* $r(\tau)$. *To use the theorem, we must calculate the covariances between* Y_s *and* Y_t, *i.e.,*

$$C[(X_s - m)(X_{s+\tau} - m), (X_t - m)(X_{t+\tau} - m)].$$

Remark 2.3. *Why divide by n instead of* $n - \tau$? *The estimator* $\widehat{r}_n(\tau)$ *is only asymptotically unbiased for large n. There are many good reasons to use the biased form in (2.16), with the n in the denominator, instead of* $n - \tau$. *The most important reason is that* $\widehat{r}_n(\tau)$ *has all the properties of a true covariance function; in Theorem 4.4 in Chapter 4, we shall see what these are.*

Example 2.15. It can be hard to interpret an estimated covariance function, and it is easy to be misled by a visual inspection. It turns out that, for large n, both the variance $V[\widehat{r}_n(\tau)]$ and the covariances $C[\widehat{r}_n(s), \widehat{r}_n(t)]$ are of the order $1/n$, and that hence the correlation coefficient $\dfrac{C[\widehat{r}_n(s), \widehat{r}_n(t)]}{\sqrt{V[\widehat{r}_n(s)] V[\widehat{r}_n(t)]}}$, between the estimated covariance function at two time points t and s, will not tend to 0 as n increases, even if both $r(s)$ and $r(t)$ are 0. Thus, there always remains some correlation between $\widehat{r}_n(s)$ and $\widehat{r}_n(t)$, regardless of how big n is. This can give the covariance function $\widehat{r}_n$ a regular and seemingly non-trivial structure even when the true correlations are quite close to 0.

The phenomenon is clear in Figure 2.15. We generated three realizations ($n = 128$) of an AR(2)-process, treated in Example 7.2, and produced three

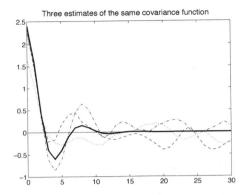

Figure 2.15 *Dashed lines: estimates of the covariance function for an AR(2)-process* $X_t = X_{t-1} - 0.5X_{t-2} + e_t$, ($n = 128$ observations). *Bold line: the true covariance function.*

estimates of its covariance function based on the observations. Note, for example, that for $\tau \approx 8 - 10$ two of the covariance estimates are clearly positive, and the third is negative, while the true covariance is almost zero.   ▲

Theorem 2.6. *(a) If* $\{X_n, n = 1, 2, \ldots\}$ *is stationary and Gaussian with mean m and covariance function* $r(\tau)$, *such that* $\sum_0^\infty r(t)^2 < \infty$, *then* $\widehat{r}_n(t)$ *defined by (2.16) is a consistent estimator of* $r(\tau)$.

(b) Under the same conditions, for $s, t = s + \tau$, *when* $n \to \infty$,

$$nC[\widehat{r}_n(s), \widehat{r}_n(t)] \to \sum_{u=-\infty}^{\infty} \{r(u)r(u+\tau) + r(u-s)r(u+t)\}. \quad (2.17)$$

(c) If $X_n = \sum_{k=-\infty}^{\infty} c_k e_{n-k}$ *is a moving average of independent, identically distributed random variables with* $E[e_k] = 0$, $V[e_k] = \sigma^2$, *and* $E[e_k^4] = \eta\sigma^4 < \infty$, *and with* $\sum_{k=-\infty}^{\infty} |c_k| < \infty$, *then the conclusion of (b) still holds, when the right hand side in (2.17) is replaced by*

$$(\eta - 3)r(s)r(t) + \sum_{u=-\infty}^{\infty} \{r(u)r(u+\tau) + r(u-s)r(u+t)\}.$$

Note that $\eta = 3$ *when* e_k *are Gaussian.*

(d) Under the conditions in (a) or (c), the estimates are asymptotically normal when $n \to \infty$.

Proof. We prove (b). Part (a) follows from (b) and Theorem 2.5. For part (c) and (d), we refer to [11, Ch. 7].

We can assume $m = 0$, and compute

$$n \, \mathsf{C}[r_n^*(s), r_n^*(t)] = \frac{1}{n} \mathsf{C} \left[\sum_{j=1}^{n-s} X_j X_{j+s}, \sum_{k=1}^{n-t} X_k X_{k+t} \right]$$

$$= \frac{1}{n} \left\{ \sum_{j=1}^{n-s} \sum_{k=1}^{n-t} \mathsf{E}[X_j X_{j+s} X_k X_{k+t}] - \sum_{j=1}^{n-s} \mathsf{E}[X_j X_{j+s}] \cdot \sum_{k=1}^{n+t} \mathsf{E}[X_k X_{k+t}] \right\}. \quad (2.18)$$

Now, it is a nice property of the normal distribution, known as Isserlis' theorem [27], that the higher product moments can be expressed in terms of the covariances. In this case,

$$\mathsf{E}[X_j X_{j+s} X_k X_{k+t}] = \mathsf{E}[X_j X_{j+s}] \mathsf{E}[X_k X_{k+t}] + \mathsf{E}[X_j X_k] \mathsf{E}[X_{j+s} X_{k+t}]$$
$$+ \mathsf{E}[X_j X_{k+t}] \mathsf{E}[X_{j+s} X_k].$$

Collecting terms with $k - j = u$ and summing over u, the normalized covariance (2.18) can, for $\tau \geq 0$, be written as

$$\sum_{u=-n+s+1}^{n-t-1} \left(1 - \frac{a(u)}{n}\right) \{r(u)r(u+t-s) + r(u-s)r(u+t)\}, \quad (2.19)$$

where

$$a(u) = \begin{cases} t + |u|, & \text{for } u < 0, \\ t, & \text{for } 0 \leq u \leq \tau, \\ s + |u|, & \text{for } u > \tau. \end{cases}$$

When $n \to \infty$ the sum (2.19) tends to $\sum_{u=-\infty}^{\infty} \{r(u)r(u+\tau) + r(u-s)r(u+t)\}$. The convergence holds under the condition that $\sum_0^\infty r(t)^2$ is finite; cf. the proof of Theorem 2.4(b). ∎

If the mean value m is unknown, one just replaces it by $\bar{x}$, the total average of the n observations $x_1, \ldots, x_n$, and instead uses

$$\widehat{r}_n(\tau) = \frac{1}{n} \sum_{t=1}^{n-\tau} (x_t - \bar{x})(x_{t+\tau} - \bar{x}),$$

as an estimate of $r(\tau)$. The conclusions of Theorem 2.6 remain true.

Example 2.16 ("Interest rates"). Figure 2.16 shows the monthly interest rates for U.S. "1-year Treasury constant maturity" bond over the 19 years 1990–2008. There seems to be a cyclic variation around a downward trend. If we remove the linear trend by subtracting a linear regression line, we get a series of data that we regard as a realization of a stationary sequence for which we estimate the covariance function. ▲

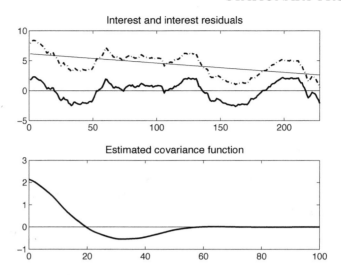

Figure 2.16 *Interest rate with linear trend, the residuals = variation around the trend line, and the estimated covariance function for the residuals.*

Testing zero correlation

In time series analysis, it is often important to be able to make a statistical test to see if variables in a stationary sequence are uncorrelated. For this one can estimate the correlation function $\rho(\tau) = r(\tau)/r(0)$ by

$$\widehat{\rho}(\tau) = \widehat{r}_n(\tau)/\widehat{r}_n(0),$$

based on n consecutive observations of the variables in the time series. If $\rho(\tau) = 0$, for $\tau = 1, 2, \ldots, p$, then the *Box-Ljung* statistic,

$$Q = n(n+2) \sum_1^p \frac{\widehat{\rho}(\tau)^2}{n - \tau},$$

has an approximate χ^2-distribution with p degrees of freedom, if n is large. This means that if $Q > \chi_\alpha^2(p)$, one can reject the hypothesis of zero correlation ($\chi_\alpha^2(p)$ is the upper α-quantile of the χ^2-distribution function, with p degrees of freedom.) The somewhat simpler *Box-Pierce* statistic,

$$Q = n \sum_1^p \widehat{\rho}(\tau)^2,$$

has the same asymptotic distribution, but requires larger sample sizes to work properly. For a discussion of limitations and alternative methods, see [11, Sec. 9.4].

2.5.4   Ergodicity a second time

Linear ergodicity means that one can estimate the expectation (= ensemble average) of a stationary process by means of the observed time average in a single realization. We have also mentioned that under certain conditions, also the covariance function, which is the ensemble average of a cross product,

$$r(\tau) = E[(X(t) - m)(X(t + \tau) - m)],$$

can be consistently estimated from the corresponding time average. A process with this property could be called *ergodic of second order*.

In a *completely ergodic* stochastic process, which we for simplicity often just call "ergodic," one can consistently estimate any expectation (= ensemble average)

$$E[g(X_{t_1}, \ldots, X_{t_p})],$$

where g is an arbitrary function of a finite number of $X(t)$-variables, by the corresponding time average in a single realization,

$$\frac{1}{n} \sum_{t=1}^{n} g(x(t + t_1), \ldots, x(t + t_p)).$$

A rigorous account of ergodic theory from a statistical viewpoint can be found in [33, Ch. 6].

Example 2.17. A histogram is usually based on data from an experiment repeated many times. In Figure 1.18, we showed a histogram over data $x_1, \ldots, x_n$, sampled from one single series of wave data. If we regard this as a realization of a (strictly) stationary process, $\{X_t\}$, we would like to interpret the histogram as an histogram of the common distribution of the X_t:s.

For an ergodic process, this interpretation is correct. The histogram is based on the relative number of observations x_t that fall in different intervals $(a_k, a_{k+1}]$. Defining $\varepsilon_t(k) = 1$, if $x_t \in (a_k, a_{k+1}]$, and $\varepsilon_t(k) = 0$ otherwise, the sequence $\{\varepsilon_t\}$ is linearly ergodic if $\{X(t)\}$ is ergodic, and hence,

$$\frac{1}{n} \sum_{t=1}^{n} \varepsilon_t(k) \to E[\varepsilon_0(k)] = P(X_0 \in (a_k, a_{k+1}]) \text{ as } n \to \infty. \qquad \blacktriangle$$

Ergodic theory is an important topic in probability theory and in physics. It helps with the interpretation of many statistical procedures, as in the previous example. The general conditions for ergodicity are of theoretical nature, but there is one simple criterion that is easy to check. For a normal stationary process one can show the following sufficient criterion; see for example [16, Sec. 7.11].

A stationary normal sequence $\{X_t, t = 1, 2, \ldots\}$ is ergodic if its co-variance function $r(\tau)$ satisfies $\frac{1}{n} \sum_{t=1}^{n} r(t)^2 \to 0$ as $n \to \infty$.

2.5.5   Processes with continuous time

For a stationary process $\{X(t), 0 \leq t < \infty\}$ with continuous time, the mean and covariance functions can be estimated in the same way as for sequences, with sums just replaced by integrals. If $X(t)$ is observed in $[0, T]$, the estimates of the mean m and the covariance function are the time averages,

$$\widehat{m}_T = \frac{1}{T} \int_0^T x(t)\, dt,$$

$$\widehat{r}_T(\tau) = \frac{1}{T} \int_0^{T-\tau} (x(t) - m)(x(t + \tau) - m)\, dt, \quad (\tau \geq 0),$$

respectively. As in the discrete case, m can be replaced by $\widehat{m}_T$ in the formula for the estimate $\widehat{r}_T(\tau)$ of the covariance function, if the mean is unknown.

The definition of ergodicity and the condition for a normal process to be ergodic are obvious integral analogies to the discrete case. If $m = 0$, and the process is Gaussian and ergodic of second order, then

$$\widehat{r}_T(0) = \frac{1}{T} \int_0^T x(t)^2\, dt \to r(0) \quad \text{as} \quad T \to \infty.$$

2.6   Stationary processes and the non-stationary reality

In the first two chapters of this book we have presented many examples of random and partly unpredictable phenomena where a stationary stochastic process is a suitable model. The characteristic feature of stationarity is that the statistical law does not change and that one therefore can use the model to predict the likely variability in the future.

However, most phenomena in the examples are not stationary in the long run. Ocean waves (Examples 1.10 and 1.14) depend on the weather; EEG recordings (Example 2.6) have different characters depending on the person's activity; stock market prices (Example 1.5) fluctuate, even the variability ("volatility") changes, sometimes in a dramatic way which is not predicted by stationary models.

So, why construct a *stationary* process model for such "unstationary" phenomena? We believe that there are two main reasons for this:

(a) *Even if the situation which is to be modeled is non-stationary in the long run, a stationary process model often gives a good description of "short-time" behavior.* It can summarize essential current information, in the examples about the ocean conditions over maybe a few hours (and after that one can perhaps change to another stationary model), about the brain activity under some specific conditions, about the current economic climate, etc. The summary of information consists of the (current) structure of the model and of the (current) parameter values. In the examples the stationary stochastic analysis is very widely used to answer questions about marine safety, health status of a patient, or the risk involved in an economic investment.

The stationary process models describe not only individual values but also dynamical properties, how values change over time, and they are therefore needed if one wants to study this, and in particular if one wants to predict future behavior.

(b) *Often stationary stochastic processes are important components in non-stationary models.* For example, it is common to model variables in the economy as a deterministic trend plus some residuals, which then are supposed to be a stationary stochastic process. As another example, a simple model for climate change could be that the average annual earth temperature consists of a linear trend plus variations around the trend, where the variations are modeled as a stationary stochastic process. Of course, stationary stochastic processes often also serve as building blocks for more complex models, which, for example, could contain cyclical trends, or trends in both mean value and variation, and perhaps also in other parts of the model.

2.7   Monte Carlo simulation from covariance function

2.7.1   Why simulate a stochastic process?

There are many good reasons why one may want to simulate samples from a stochastic process, and we list some of them here.

- The first reason is purely educational: to learn about and illustrate the behavior of a stochastic model, just as is done in this book.

- An important practical reason is the need for test data in experiments with new inventions or constructions. Before a physical test is made on a mechanical device extensive computer simulations are always made where a realistic environment is generated as a stochastic process in time and/or space. Depending on the type of construction, different aspects of the process are then important. In a vibration test it may be the frequency content of the signal that is most important for the resonant behavior. In a fatigue test of a mechanical part in a car engine it is the sequence of maxima and minima that has to be representative for the working environment. As a third example one may take an image or signal detection system where one has to generate both possible variation in the signal and the randomness of the environment.

- A reason of statistical importance is the desire to evaluate statistical procedures like estimation or testing methods, in particular when data are correlated in time. An example is testing of a process model with parameters estimated from a time series. Repeated simulations are generated from the fitted model and new estimates calculated. The observed variation in the estimates can be used to assess the confidence in the model; this procedure is called *parametric bootstrap*.

It is obvious from the examples that there is a need for many different simulation techniques, mainly depending on the model information at hand and on the desired length of the generated data. For stationary processes there are three main techniques.

From covariance: This technique is suitable for Gaussian processes since these are completely determined by mean value and covariance structure.

From spectrum: This is a very flexible technique and can be used to generate almost any type of dependence in a Gaussian stationary process; see Section 5.6.

From ARMA-structure: This technique has the advantages that it easily can handle also non-Gaussian processes, and that it makes it easy to add new observations to an already simulated time series; see Section 7.6.

2.7.2 Simulation of Gaussian process from covariance function

Let $\boldsymbol{\mu} = \mathsf{E}[\mathbf{X}]$ be the mean value vector and $\boldsymbol{\Sigma} = (\sigma_{jk})$ the covariance matrix of a vector of normal random variables $\mathbf{X} = (X_1, \ldots, X_p)'$. Observations of $\mathbf{X}$ can then be generated from p independent standard normal variables $\mathbf{U} = (U_1, \ldots, U_p)'$. Independent normal variables in turn are very easy to generate. We now briefly describe two methods for doing such simulations.

Cholesky factorization

Assume zero means and note that $\boldsymbol{\Sigma}$ is symmetric and positive definite. Then the *Cholesky factorization* of $\boldsymbol{\Sigma}$ is a representation

$$\boldsymbol{\Sigma} = \mathbf{L}\mathbf{L}'$$

with a lower triangular matrix $\mathbf{L}$, ($c_{jk} = 0$ for $k > j$). When $\mathbf{U} = (U_1, \ldots, U_p)'$ consists of independent standard normal variables, it is easily seen that

$$\mathbf{X} = \mathbf{L}\mathbf{U}$$

has covariance matrix $\boldsymbol{\Sigma}$.

Thus, to simulate a sample $\mathbf{x}(n) = (x_1, \ldots, x_n)'$ from a stationary or non-stationary zero mean Gaussian process one follows this simple scheme.

1. Create the covariance matrix $\boldsymbol{\Sigma}(n)$ from the covariance function,
2. Find the Cholesky factorization $\mathbf{L}(n)\mathbf{L}(n)' = \boldsymbol{\Sigma}(n)$,
3. Generate independent standard normal $\mathbf{u}(n) = (u_1, \ldots, u_n)'$,
4. Take $\mathbf{x}(n) = \mathbf{L}(n)\mathbf{u}(n) + \boldsymbol{\mu}$,
5. To add a new time point, update $\mathbf{L}(n)$ by adding a new row L_{n+1}, generate u_{n+1}, and compute $x(n+1) = L_{n+1}(\mathbf{u}(n)' u_{n+1})'$.

The update in step 5 is numerically much easier than one would think at first glance: one doesn't have to redo all the previous steps from the beginning; see [2, Ch. XI.2].

Circulant embedding

A circulant covariance matrix has the very special structure $C[X(j), X(k)] = C[X(j+\tau), X(k+\tau)]$. The covariance matrix of a stationary process can be embedded in a *circulant matrix* as in the following example,

$$\boldsymbol{\Sigma}(3) = \begin{pmatrix} r(0) & r(1) & r(2) \\ r(1) & r(0) & r(1) \\ r(2) & r(1) & r(0) \end{pmatrix} \quad \text{is embedded in} \quad \mathbf{C} = \begin{pmatrix} r(0) & r(1) & r(2) & r(1) \\ r(1) & r(0) & r(1) & r(2) \\ r(2) & r(1) & r(0) & r(1) \\ r(1) & r(2) & r(1) & r(0) \end{pmatrix}.$$

(There is no guarantee that this in fact is a covariance matrix. However, there are methods to overcome this difficulty.) A circulant matrix has a particularly simple eigenvector decomposition. This makes simulation much more efficient than for the Cholesky decomposition; for details, see [2, Ch. XI.3]. However, on the other hand, for this simulation method, updating to add one more observation is more difficult than for the Cholesky factorization method: it requires redoing the entire computation from the beginning.

Exercises

2:1. Let X, Y, and Z be independent random variables with expectation 1 and variance 1. Compute

(a) $E[3X - 2Y]$,   (b) $E[aX + bY]$,   (c) $V[3X - 2Y]$,

(d) $V[aX + bY]$,   (e) $V[aX - bY]$,   (f) $C[3X - 2Y, X + Y - 2Z]$.

2:2. What is the correlation coefficient, $\rho[X_1, X_2]$, between the random variables X_1 and X_2 if $V[X_1] = 5$, $V[X_2] = 10$, and $C[X_1, X_2] = -4.5$?

2:3. Let $e_0, e_1, \ldots$ be a sequence of independent, identically distributed random variables with mean m and variance σ^2. Let $\{X_t, t = 1, \ldots\}$ be the stochastic process defined by $X_t = 1.2e_t + 0.9e_{t-1}$. Compute the mean $m_X(t) = E[X(t)]$, and the covariance function $r_X(s,t) = C[X_s, X_t]$. Show that $\{X_t\}$ is weakly stationary.

2:4. Let $X_1, X_2, \ldots$ be a sequence of zero-mean, independent, identically distributed random variables with variance σ^2. Define

$$Y_1 = X_1,$$
$$Y_k = 0.5Y_{k-1} + X_k, \quad k \geq 2.$$

Prove that $\{Y_k\}$ is non-stationary by computing $r_Y(1,2)$ and $r_Y(2,3)$. The non-stationarity can be corrected by letting $Y_1 = cX_1$ for an appropriate choice of the constant c. Find such a c.

2:5. Find a physically realistic explanation to the dependence structure at lag $\tau = 9$ in Figure 2.3.

2:6. Let Y and Z be two independent random variables with expected values 0 for both and variances σ_y^2 and σ_z^2. A stochastic process is created from

$$X(t) = Y \sin(t) + Z + 5.$$

Is the process weakly stationary?

2:7. Complete the "easy to see" part of the convergence in (2.8).

2:8. A physicist studies the amount of carbon dioxide, $W(t)$, at time t, near a highly trafficked road and models the amount by a stationary process. During one day she gets the measurements $w(1), w(2), \ldots, w(50)$, which she regards as a realization of the stationary stochastic process $W(t)$. In order to study the dependence between $W(t)$, $W(t+1)$, and $W(t+2)$ she makes two scatter plots at time lags 1 and 2 of the observed data, that is, she plots the points $(w(t), w(t+1))$ and $(w(t), w(t+2))$; see Figure 2.17. What can you tell from them? She has to choose amongst the following two different stochastic processes:

Option 1: $W(t) = e(t) + e(t-1) + \dfrac{1}{3}e(t-2) - \dfrac{1}{3}e(t-3)$,

Option 2: $W(t) = e(t) - e(t-1) + \dfrac{1}{3}e(t-2) + \dfrac{1}{3}e(t-3)$,

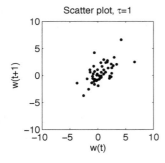
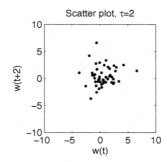

Figure 2.17 *Scatter plots of pairs of measurements in Exercise 2:8.*

where $e(t)$ are zero-mean, independent, identically distributed random variables with variance 1. Which one should she choose? Hint: Consider the covariance functions for the two options:

Option 1:

$$r_W(\tau) = \begin{cases} 20/9, & \tau = 0, \\ 11/9, & \tau = \pm 1, \\ 0, & \tau = \pm 2, \\ -1/3, & \tau = \pm 3, \\ 0, & \text{otherwise.} \end{cases}$$

Option 2:

$$r_W(\tau) = \begin{cases} 20/9, & \tau = 0, \\ -11/9, & \tau = \pm 1, \\ 0, & \tau = \pm 2, \\ 1/3, & \tau = \pm 3, \\ 0, & \text{otherwise.} \end{cases}$$

2:9. Which of the following functions can be the covariance functions to a stationary process and which can not?

(a) $r(\tau) = \frac{\sin \tau}{\tau}$,   (b) $r(\tau) = \frac{\sin \tau}{\tau^2}$,   (c) $r(\tau) = \begin{cases} 1, & \tau = 0, \\ 2, & \tau = 1, \\ 0.5, & \tau = 2, \\ 0, & \text{otherwise.} \end{cases}$

(d) $r(\tau) = \cos \tau$,   (e) $r(\tau) = e^{|\tau|}$.

2:10. A digital image consists of pixels, $M(x,y)$, arranged in a large matrix. Each pixel takes a real value, which determines its color. The image captured by a digital camera is often affected by noise if the light is dim. In order to reduce the effect of the noise we replace each pixel value of the image with the average of the original pixel value and the four closest pixel values:

$$\widehat{M}(x,y) = \frac{M(x,y)+M(x-1,y)+M(x+1,y)+M(x,y-1)+M(x,y+1)}{5}.$$

Using statistical terminology, we might say that we estimate the "true" pixel value with the average of the outcomes of five (possibly correlated) random variables. Your task is to compute the variance of this estimate. We consider the image as a stochastic process:

$$M(x,y) = S(x,y) + N(x,y),$$

where $S(x,y)$ is a constant (the "true" image) and $N(x,y)$ is correlated noise with covariance $C[N(x_1,y_1),N(x_2,y_2)] = \rho^\delta \sigma^2$, where $\rho = 0.5$, $\sigma = 1$, and δ is the ordinary euclidian distance between the pixel (x_1,y_1) and (x_2,y_2). Using this model, compute the variance $V[\widehat{M}(x,y)]$.

2:11. Let Y be a random variable such that $P(Y = -1) = P(Y = 1) = 0.5$. Let Z be a random variable uniformly distributed in the interval $[0, 2\pi)$.

(a) Is $X(t) = Y\sin(t), t \in \mathbb{R}$, a weakly stationary process?

(b) Is $W(t) = \sin(t+Z)$, $t \in \mathbb{R}$, a weakly stationary process?

2:12. A weakly stationary process has expected value m and covariance function $r(0) = 3$, $r(\pm 1) = -1$ and zero for all other values.

a) Compute, for $N = 15$, the standard deviation of the estimate $\overline{X} = \frac{1}{N}\sum_{k=1}^{N} X_k$.

b) If we have the possibility to choose the time between the measure points in a), i.e., $\overline{X} = \frac{1}{N}\sum_{k=1}^{N} X_{\alpha k}$, how should the integer α be chosen for lowest possible standard deviation?

2:13. We are interested in a low frequency signal, but unfortunately our bias-free measurements of the signal $\{X(n)\}$ are disturbed by noise. We model our recorded signal as a deterministic low frequency component (the one we are interested in) plus a stationary stochastic process (the noise). We have reasons to believe that the covariance function, $r(\tau)$, of the noise can be approximated with

$$r(\tau) = \begin{cases} 1.2 & \tau = 0, \\ 0.5 & \tau = \pm 1, \\ 0 & \text{otherwise.} \end{cases}$$

In order to suppress the noise, we consider the following two options:

$$Y_N(n) = \frac{1}{N}\underbrace{\left(X(n) + X(n-1) + \cdots + X(n-N+1)\right)}_{N \text{ terms}},$$

$$Z_M(n) = \frac{1}{M}\underbrace{\left(X(n) + X(n-2) + \cdots + X(n-2(M-1))\right)}_{M \text{ terms}},$$

where $\{Y_N(n)\}$ and $\{Z_M(n)\}$ are two different approximations of the low frequency signal of our interest.

(a) Why are $\{Y_N(n)\}$ and $\{Z_M(n)\}$ less affected by noise than our original measurements, $\{X(n)\}$?

(b) There is a trade-off in selecting N (or M). Explain!

(c) How large must we choose N in order for the variance of $Y_N(n)$ to be less than 0.1? How large must we choose M in order for the variance of $Z_M(n)$ to be less than 0.1?

(d) Compare the values of N and M in the previous question. Give an explanation of the difference.

2:14. A division of the national Swedish Institute for Building Research wants to estimate the average noise level in a specific area. The dependence between measurements, $\{Y_j\}$, at different locations is described by $C[Y_i, Y_j] = r_Y(d) = e^{-d}$, where d is the distance between the i:th and the j:th measurement point. They are restricted to estimate the noise level with the average of the sound level recorded by three microphones. They discuss two different placements, shown in Figure 2.18. Do some computations, and argue in favor of one of the alternatives.

2:15. A zero-mean sequence, $\{X_k\}$, has covariance function

$$r(\tau) = 2^{-|\tau|/10} \quad \tau = 0, 1, \ldots .$$

We would like to estimate m by observing T consecutive samples of the process. How large must T be, in order for the standard deviation of the estimate to be less than 0.1? You can make appropriate approximations.

2:16. Suppose that $\{X_k\}$ is a stationary process with unknown mean m, known variance σ^2, and correlation function $r(\tau) = 0.5^{|\tau|}$. The mean m is estimated with $\frac{1}{N}\sum_1^N X_k$. Suppose that N is large and approximate the variance of the estimator. Also, find the number of elements that would have been necessary in order to achieve the same variance, if the elements of $\{X_k\}$ had been uncorrelated.

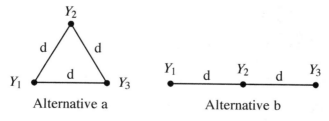

Figure 2.18 *Two different microphone placements.*

2:17. In Example 2.12 we presented a general rule of thumb for how many observations are needed to get the same variance as for independent variables in an estimate of the mean. Motivate the rule and show that it is exact if the correlation function is of the form $\rho(\tau) = \theta^{|\tau|}$.

2:18. The Swedish Meteorological and Hydrological Institute (SMHI) reports daily ozone levels for a number of Swedish cities. We are particularly interested in the measurements during July, which seems like an "ordinary" month, with mean 335 and standard deviation 16. We assume that the observations, $x_1, \ldots, x_{31}$, come from a stationary Gaussian process, $\{X_t\}$. The covariance function is stated below for a few time lags,

τ:	0	1	2	15	30
$r(\tau)$:	256	163	159	-75	-9

Suppose that we want to estimate the mean value, using only three measurements during July. Which measurements should we choose? Answer the question by computing the variance of

$$m_1 = \frac{X_1 + X_{16} + X_{31}}{3} \quad \text{and} \quad m_2 = \frac{X_{15} + X_{16} + X_{17}}{3}.$$

2:19. Repeat Example 2.11 with the model $X_{t+1} = -0.9X_t + e_t$ to see how negative correlation affects the estimate.

2:20. Let Y and $\{X_n, n = 1, 2, \ldots\}$ be independent normal variables and define $Z_n = X_n + Y$. Is $\{Z_n, n = 1, 2, \ldots\}$ a linearly ergodic sequence? Compare with the covariance criterion for ergodicity.

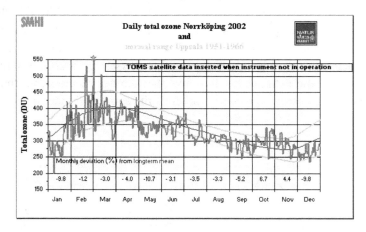

Figure 2.19 *Ozone levels in Norrköping used in Exercise 2:18, (SMHI).*

Chapter 3

The Poisson process and its relatives

3.1 Introduction

The Poisson[1] process plays a special role among continuous time processes with discrete state space. It is used as a model for experiments where events of interest occur at independent random time points with constant average rate. For instance, the number of telephone calls, visits to web pages, vehicles passing an intersection in a road network with light traffic, car accidents, emissions from a radioactive isotope can be modeled by Poisson processes, provided conditions are stationary and there is no interaction between events. The Poisson process is the simplest example of a stationary *point process*.

The chapter presents three equivalent characterizations of the Poisson process, by its counting properties, by inter-arrival times, and by uniform conditional independent arrivals. We discuss methods to describe dependence between events, and Poisson processes in the plane. The chapter concludes with an overview of Monte Carlo simulation of various types of Poisson processes.

3.2 The Poisson process

Counting the number of events from some starting time point by letting

$$X(t) = \text{number of events in the time interval } (0, t],$$

the variable $X(t)$ increases by 1 each time an event occurs. If more than one event can occur at the same time, $X(t)$ can increase by several units at a time. Figure 3.1 shows the relation between the event times – marked by dots on the horizontal axis – and the counting process $X(t)$. The Poisson process $X(t)$ is, of course, not a stationary process, since its mean value increases with t. However, it belongs to the stationary family, since it counts a stream of events that has stationary statistical properties.

---

[1] Siméon Denis Poisson, French mathematician, physicist, and probabilist, 1781–1840.

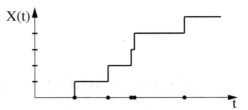

Figure 3.1 *Counting events in the interval* $(0,t]$.

3.2.1   *Definition and simple properties*

There are many equivalent definitions of a Poisson process. We start with the two most simple ones: the counting definition and the arrival time definition. The increment of a stochastic process $X(t)$ over an interval, say the interval $(t, t+h]$, is the difference $X(t+h) - X(t)$ of the value of the process at the end of the interval and the beginning of the interval. A process has *independent increments* if the increments over any collection of disjoint intervals are statistically independent.

Counting definition

Definition 3.1 (First definition of a Poisson process). *A stochastic process $\{X(t), 0 \le t < \infty\}$ is a Poisson process with intensity $\lambda > 0$ if $X(0) = 0$, and it has independent increments such that $X(t+h) - X(t)$, for any $t, h > 0$, has a Poisson distribution with mean λh, so that*

$$P(X(t+h) - X(t) = k) = e^{-\lambda h} \frac{(\lambda h)^k}{k!} \quad \text{for} \quad k = 0, 1, \dots, \quad (3.1)$$

$$\mathsf{E}[X(t)] = \mathsf{V}[X(t)] = \lambda t, \quad (3.2)$$

$$\mathsf{E}\left[\frac{X(t+h) - X(t)}{h}\right] = \lambda. \quad (3.3)$$

From (3.1) we get the following important properties and interpretation:

$$\frac{P(X(t+h) - X(t) = 1)}{h} = \lambda e^{-\lambda h} \to \lambda \quad \text{as } h \downarrow 0, \quad (3.4)$$

$$\frac{P(X(t+h) - X(t) > 1)}{h} = h^{-1} e^{-\lambda h} \sum_{k=2}^{\infty} \frac{(\lambda h)^k}{k!} \to 0 \quad \text{as } h \downarrow 0. \quad (3.5)$$

Thus, by (3.4), the probability of exactly one event in a small interval is approximately proportional to the interval length, and, by (3.5), the probability of more than one event is of smaller order. Thus, in a Poisson process, there is zero probability that two events happen at exactly the same time.

In a Poisson process the mean value of $X(t)$ is equal to the variance; by (3.2) both are equal to λt. Note that λ is not a probability – it can be greater than one – but rather a "probability per time unit."

Example 3.1 ("Marine accidents in Swedish waters"). Accidents or "near accidents" with Swedish merchant and fishing vessels occur at an average rate of 2.5 per week, taken as an average over the whole year. Assuming accidents occur according to a Poisson process with intensity $\lambda = 2.5$ [week^{-1}], the standard deviation of the number of accidents in a week is $\sqrt{2.5} = 1.58$. The expected number of accidents over four weeks is ten, with a standard deviation $\sqrt{10} = 3.16$. The standard deviation doubles with a four times as long time interval, and the coefficient of variation (= relative standard deviation = standard deviation divided by expected value) reduces by a factor of two. ▲

Example 3.2 ("Traffic deaths"). Figure 3.2 shows the number of traffic deaths on roads in Southern Sweden during the years 1997 to 2009; on average 66 deaths per year. The downward trend in traffic deaths during the period is statistically significant on the 5% level.

An issue of great public interest is if a single year with increased deaths indicates that the favorable trend is broken or not. If accidents occur according to a Poisson process one can assume it to have intensity $\lambda = 66$ [year^{-1}], and hence the standard deviation of the yearly numbers is around $8 \approx \sqrt{66}$. As seen, the fluctuations around the trend are in agreement with that number.

There are good reasons against the use of the Poisson process as a model for traffic deaths – many accidents lead to more than one casualty, which is not possible in the Poisson process model. If we, regardless of this objection, use the Poisson process to find the probabilities $p_0, p_1, p_2, \ldots$ for $0, 1, 2, \ldots$ deaths during a single week, we get the following table. It also shows the average number of weeks per year (52 weeks) $= 52\, p_k$, with k deaths. As seen in the table, even if the average number of casualties per week is slightly more than one, one can expect several weeks with $3, 4$, and even more deaths.

k	0	1	2	3	≥ 4
$p_k = e^{-66/52}\, \dfrac{(66/52)^k}{k!}$	0.281	0.357	0.226	0.096	0.040
$52\, p_k$	14	19	12	5	2

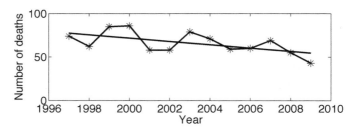

Figure 3.2 *Broken line: The yearly number of traffic casualties in Southern Sweden 1997 to 2009. Straight line: estimated trend line. The Poisson variation around the average number (66 deaths per year) between years has a standard deviation of about $8 \approx \sqrt{66}$, so one can expect large variations from year to year.*

A better model for the traffic accident statistics would be a process where the number of accidents, not the number of deaths, follows a Poisson process. An even more realistic model would be to let the intensity depend on time, for example to account for a possible trend; see Section 3.6. For more details, see [24] or [25].   ▲

Example 3.3 ("Estimation of intensity"). Estimation of the intensity λ in a Poisson process is simple: if n_T events have been observed during a time interval of length T, then $\widehat{\lambda}_T = n_T/T$ is an unbiased estimate of λ. To find the uncertainty of the estimate is almost as simple. Since n_T is an observation of a Poisson variable with mean λT, its variance is also λT; this is a basic property of the Poisson distribution. Then the standard deviation of the estimate $\widehat{\lambda}_T$ is $\mathsf{D}[\widehat{\lambda}_T] = \sqrt{\lambda/T}$. An estimate of $\mathsf{D}[\widehat{\lambda}_T]$ is given by $d[\widehat{\lambda}_T] = \sqrt{\widehat{\lambda}_T/T}$.

Further, if λT is not too small, the Poisson distribution can be approximated by a normal distribution; this follows from the central limit theorem. Following the scheme in Example 2.14 we can therefore construct a confidence interval for λ with confidence level $1 - \alpha$, as

$$I_\lambda: \quad \{\widehat{\lambda}_T - q_{\alpha/2}\mathsf{d}[\widehat{\lambda}_T], \widehat{\lambda}_T + q_{\alpha/2}\mathsf{d}[\widehat{\lambda}_T]\}. \qquad ▲$$

3.2.2   Interarrival time properties

Our first definition of the Poisson process was based on the number of events occurring in disjoint intervals. Our next definition will directly model the times of the different events, by specifying the distribution of the time that elapses between successive events.

Now, let $\{X(t), t \geq 0\}$ be a Poisson process according to Definition 3.1 and denote by T_1 the time for the first event. Obviously, T_1 is greater than

t if and denote no event occurs in the interval $[0,t]$. We conclude, from the Poisson probability, that T_1 has an exponential distribution:

$$P(T_1 > t) = P(X(t) = 0) = e^{-\lambda t}.$$

This simple statement can be generalized to a characteristic property of the Poisson process, which we state without proof; for a proof, see [25, Chapter VII]. We let $T_1, T_2, \ldots$ be the time intervals between the points of increase in $X(t)$, so the events occur at times $T_1, T_1 + T_2, \ldots$.

> **Theorem 3.1.** (Alternative characterization of a Poisson process: Independent exponential interarrival times). *A process $\{X(t), 0 \leq t < \infty\}$ is a Poisson process with intensity λ, if and only if the interarrival times $T_1, T_2, \ldots$, are independent and exponentially distributed with mean $1/\lambda$.*

As stated in the theorem, independent, exponential interarrival times is a sufficient property for a counting process to be a Poisson process, under the condition that the process has jumps only of size one. Therefore, the property in the theorem can be used as a second definition of a Poisson process.

The theorem states that the waiting time from one event, for example, event number n, to the next event is independent of all previous waiting times. As a consequence, the waiting time is independent of the time $\tau_n = T_1 + T_2 + \ldots + T_n$ at which the nth event occurs. Now, as is well known, the exponential distribution "lacks memory," which means that if the random waiting time T is exponential with mean $1/\lambda$, then, for $\tau > 0$,

$$P(T > t_0 + \tau \mid T > t_0) = \frac{P(T > t_0 + \tau)}{P(T > t_0)} = \frac{e^{-\lambda(t_0+\tau)}}{e^{-\lambda t_0}} = e^{-\lambda \tau} = P(T > \tau).$$

Hence, the remaining waiting time has a distribution that does not depend on the waiting time t_0 already spent. This has the following consequence for the waiting times in a Poisson process.

> **Theorem 3.2.** *In a Poisson process $\{X(t), t \geq 0\}$ with intensity λ, the waiting time from any fixed time t_0 to the next event is exponential with mean $1/\lambda$.*

Example 3.4 ("Traffic deaths"). Based upon the Swedish statistics for road accidents, let us assume that, in a certain region, accidents with deadly outcome occur according to a Poisson process with constant intensity $\lambda = 1[\text{week}^{-1}]$. What is then the expected waiting time between two such accidents, and what is the probability that it lasts at least three weeks from one accident to the next? Suppose there has been no fatal accident during all of April. What is then the probability of no such accident during the following three week period May 1–May 21 ?

Answer: By Theorem 3.1 the time between accidents has an exponential distribution with mean $1/\lambda$ and hence the expected waiting time is seven days. The probability of at least three weeks from one accident to the next is $P(T > 3) = e^{-3} = 0.05$. By Theorem 3.2 the waiting time from May 1 to the next accident does not depend on what happened in April, so the probability of no accident during the three week period is also 0.05.   ▲

3.2.3   Some distribution properties

In this section we derive an alternative characterization of the Poisson process. A Poisson process starts with $X(0) = 0$ and $X(t)$ has a Poisson distribution with mean λt, i.e., it has the one-dimensional probability function

$$p_t(x) \; = \; p_{X(t)}(x) = P(X(t) = x) = e^{-\lambda t}\frac{(\lambda t)^x}{x!} \quad \text{for} \quad x = 0, 1 \dots.$$

We will now show how the two-dimensional distribution for the Poisson process can be derived. Take two time points $0 < s < t$, and calculate the probability

$$p_{s,t}(x,y) = P(X(s) = x, X(t) = y).$$

First, we realize that the probability is 0 if $x > y$. For $x \leq y$, we will use the fact that the events

$$\{X(s) = x, \; X(t) = y\} \quad \text{and} \quad \{X(s) = x, \; X(t) - X(s) = y - x\}$$

are equivalent. Now $X(s)$ and $X(t) - X(s)$ are the number of events in the disjoint intervals $(0, s]$ and $(s, t]$, respectively, and since the Poisson process has independent increments, the variables $X(s)$ and $X(t) - X(s)$ are independent. Since they furthermore are Poisson distributed with expectation equal to λ

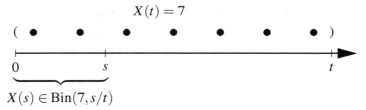

Figure 3.3 *If it is known that $X(t) = 7$, when did the events happen? Partial answer: The conditional distribution of $X(s)$ given $X(t) = 7$, $s < t$, is binomial with parameters 7 and s/t.*

times the length of the interval, we can factorize $p_{s,t}(x,y)$ as follows,

$$P(X(s) = x, X(t) = y) = P(X(s) = x, X(t) - X(s) = y - x)$$

$$= P(X(s) = x)P(X(t) - X(s) = y - x) = e^{-\lambda s} \frac{(\lambda s)^x}{x!} \cdot e^{-\lambda(t-s)} \frac{(\lambda(t-s))^{y-x}}{(y-x)!}$$

$$= e^{-\lambda t} \frac{(\lambda t)^y}{y!} \cdot \frac{y!}{x!(y-x)!} \cdot \left(\frac{s}{t}\right)^x \left(\frac{t-s}{t}\right)^{y-x}, \tag{3.6}$$

for $y \geq x \geq 0$.

The interpretation of the bivariate distribution and the nice product form (3.6) is important. If we have observed a certain number of events in a Poisson process up to time t, say $X(t) = y$, can anything be said about *when* the events happened? Yes, one may, and we now show how.

If $X(t) = y$, we calculate the conditional probability that $X(s) = x$ for $s < t$ and $x \leq y$:

$$P(X(s) = x \mid X(t) = y) = \frac{P(X(s) = x, X(t) = y)}{P(X(t) = y)}$$

$$= \frac{e^{-\lambda t} \frac{(\lambda t)^y}{y!} \cdot \frac{y!}{x!(y-x)!} \cdot \left(\frac{s}{t}\right)^x \left(\frac{t-s}{t}\right)^{y-x}}{e^{-\lambda t} \frac{(\lambda t)^y}{y!}} = \frac{y!}{x!(y-x)!} \cdot \left(\frac{s}{t}\right)^x \left(\frac{t-s}{t}\right)^{y-x}.$$

This is nothing but the Binomial probability for x successful outcomes in y independent repetitions of an experiment, with success probability s/t. We see that out of the y events that have happened before time t, there is a Binomial probability that x of these had occurred already at time s. This can be generalized, and gives the following interpretation for the Poisson process: The events that happened in the interval $(0, t]$ occurred at independent, uniformly distributed time points between 0 and t. This is the mathematical formulation of the property that "events occur at independent random time points and with uniform average rate"; for a proof see [24]. For future reference we formulate the following remark.

> **Remark 3.1** (Uniform conditional arrivals). *A Poisson process over a finite interval can be generated in a two-step random procedure: First decide on the number of events that shall occur in the interval – do that according to a Poisson distribution with mean λ times the length of the interval. Then, select the times for the events independently of each other, and uniformly over the interval.*

3.3 Stationary independent increments

3.3.1 A general class with interesting properties

The Poisson process is a special case of a class of processes that have *stationary independent increments*. Let us specify the statistical properties of a counting process for independent single events with constant event rate.

> **Definition 3.2.** *A stochastic process $\{X(t), t \in T\}$ is said to*
>
> a) **have independent increments** *if the increments $X(t_2) - X(t_1)$, $X(t_3) - X(t_2), \ldots, X(t_n) - X(t_{n-1})$ are independent for every set of times, $t_1 \leq t_2 \leq \cdots \leq t_n$ in T;*
>
> b) **have stationary increments** *if the distribution of $X(t + h) - X(t)$ does not depend on t, only on h;*
>
> c) **be simply increasing** *if it is non-decreasing with integer jumps, and*
> $$P(X(t+h) - X(t) > 1)/h \to 0 \text{ as } h \downarrow 0.$$
>
> d) **Third definition of a Poisson process:** *A simply increasing stochastic process $\{X(t), 0 \leq t < \infty\}$ with $X(0) = 0$ and with stationary, independent increments is called a homogeneous Poisson process.*

Example 3.5. Let $X(t)$ be the number of telephone calls to a telephone switch during the time interval $(0, t]$. Then, $X(1) - X(0)$ is the number of calls during the first second, $X(2) - X(1)$ the number of calls during the next second, and so on. "Independent increments" in this case means that the number of calls during different intervals are statistically independent. "Stationary increments" means that the distribution of calls does not change during the day. Finally, that the process is "simply increasing" means that only one call can arrive at a time. The reader is encouraged to consider how realistic the independence and stationarity assumptions are in this and other examples of counting processes. For example, what happens if only a small number of customers use the switchboard? And what about diurnal effects? ▲

Remark 3.2. *The Poisson process can be used to count the number of unique events that do not interact. The condition that events do not interact has, for example, the consequence that it may be used to count car accidents, but not to count the number of cars involved in accidents, since each time one car is involved in an accident, there is a positive probability that there will be more cars involved. In fact, an accident may cause a sequence of accidents, so the Poisson process may not be the ideal model, in that case either. We will present some modifications that take care of these situations, namely the* compound Poisson process *and the* shot noise *process in Section 5.5.*

3.3.2 Covariance properties

Processes with independent, stationary increments, like the Poisson process, have particularly simple expectation and covariance functions.

Let the process $\{X(t), t \geq 0\}$ start at $X(0) = 0$ and have independent, stationary increments, with finite variance. Thus, the distribution of the change $X(s+t) - X(s)$ over an interval $(s, s+t]$ only depends on the interval length t and not on its location. In particular, $X(s+t) - X(s)$ has the same distribution as $X(t) = X(t) - X(0)$, and also the same mean $m(t)$ and variance $v(t)$. We first show that both $m(t)$ and $v(t)$ are proportional to the interval length t.

Mean value function: Since $E[X(s+t) - X(s)] = E[X(t)]$, one has

$$m(s+t) = E[X(s+t)] = E[X(s)] + E[X(s+t) - X(s)] = m(s) + m(t),$$

which means that for $s, t \geq 0$, the mean function is a solution to the equation

$$m(s+t) = m(s) + m(t),$$

which is known as Cauchy's functional equation. If we now look only for continuous solutions to the equation, it is easy to argue that $m(t)$ is of the form (note: $m(0) = 0$),

$$m(t) = E[X(t)] = k_1 \cdot t, \quad t \geq 0,$$

for some constant $k_1 = m(1)$. (The reader could prove this by first taking $t = 1/q$, with integer q, and then $t = p/q$ for integer p.)

The variance function: The variance has a similar form, which follows from the independence and stationarity of the increments. For $s, t \geq 0$, we write $X(s+t)$ as the sum of two increments, $Y = X(s) = X(s) - X(0)$, and $Z = X(s+t) - X(s)$. Then, Y has variance $V[Y] = V[X(s)] = v(s)$, by definition. The second increment is over an interval of length t, and since the distribution of an increment only depends on the interval length, Z has the same distribution as $X(t)$, and hence $V[Z] = V[X(t)] = v(t)$. Thus,

$$v(s+t) = V[X(s+t)] = V[Y+Z] = V[Y] + V[Z] = v(s) + v(t).$$

As before, the only continuous solution is $v(t) = V[X(t)] = k_2 \cdot t$, for some constant $k_2 = V[X(1)] \geq 0$. Thus, we have shown that both mean and variance functions are proportional to the interval length.

The covariance function: Finally, we turn to the covariance function. First take the case $s \leq t$. Then, we can split $X(t)$ as the sum of $X(s)$ and the increment from s to t, and get

$$
\begin{aligned}
r(s,t) &= C[X(s), X(t)] = C[X(s), X(s) + (X(t) - X(s))] \\
&= C[X(s), X(s)] + 0 = V[X(s)] = k_2 \cdot s.
\end{aligned}
$$

For $s > t$, we just interchange t and s, and realize that it is the minimum time that determines the covariance: $r(s,t) = k_2 \cdot t$.

We sum up the results in a general theorem.

Theorem 3.3. a) *The mean value and covariance functions for a process with stationary independent increments starting with* $X(0) = 0$ *are*

$$
\begin{aligned}
m(t) &= E[X(t)] = E[X(1)] \cdot t, \\
r(s,t) &= V[X(1)] \cdot \min(s,t).
\end{aligned}
$$

b) *A Poisson process with intensity* λ *has the moment functions*

$$
\begin{aligned}
m(t) &= E[X(t)] = \lambda t, \\
r(t,t) &= V[X(t)] = \lambda t, \\
r(s,t) &= C[X(s), X(t)] = \lambda \min(s,t).
\end{aligned}
$$

Besides the Poisson process, we will introduce the Wiener process as an important example of a process of this type; see Section 5.3.

Example 3.6 ("Covariance function for the Poisson process"). The covariance function can also be obtained in a more cumbersome way by direct calculation from the bivariate probability distribution, derived in Section 3.2.3. For $s < t$,

$$E[X(s)X(t)] = \sum_{x,y} xy\, p_{X(s),X(t)}(x,y)$$
$$= \sum_{0 \le x \le y} xy\, e^{-\lambda t}\frac{(\lambda t)^y}{y!}\binom{y}{x}\left(\frac{s}{t}\right)^x\left(1-\frac{s}{t}\right)^{y-x} = \dots = \lambda s + \lambda^2 st,$$

giving $r(s,t) = \lambda s + \lambda^2 st - \lambda^2 st = \lambda s$, and in general $r(s,t) = \lambda \cdot \min(s,t)$, as it should. ▲

Example 3.7 ("Random telegraph signal"). This extremely simple process jumps between two states, 0 and 1, according to the following rules.

Let the signal $X(t)$ start at time $t = 0$ with equal probability for the two states, i.e., $P(X(0) = 0) = P(X(0) = 1) = 1/2$, and let the switching times be decided by a Poisson process $\{Y(t), t \ge 0\}$ with intensity λ independently of $X(0)$. Then, $\{X(t), t \ge 0\}$ is a weakly stationary process; in fact, it is also strictly stationary, but we don't show that.

Let us calculate $E[X(t)]$ and $E[X(s)X(t)]$. At time t, the signal is equal to

$$X(t) = \frac{1}{2}\left(1 - (-1)^{X(0)+Y(t)}\right),$$

since, for example, if $X(0) = 0$ and $Y(t)$ is an even number, $X(t)$ is back at 0, but if it has jumped an uneven number of times, it is at 1. Since $X(0)$ and $Y(t)$ are independent,

$$E[X(t)] = E\left[\frac{1}{2}(1 - (-1)^{X(0)+Y(t)})\right] = \frac{1}{2} - \frac{1}{2}E\left[(-1)^{X(0)}\right]E\left[(-1)^{Y(t)}\right],$$

which is constant equal to $1/2$, since $E[(-1)^{X(0)}] = \frac{1}{2}(-1)^0 + \frac{1}{2}(-1)^1 = 0$. As a byproduct, we get that $P(X(t) = 0) = P(X(t) = 1) = E[X(t)] = \frac{1}{2}$.

For $E[X(s)X(t)]$, we observe that the product $X(s)X(t)$ can be either 0 or 1, and it is 1 only when both $X(s)$ and $X(t)$ are 1. Therefore, for $s < t$,

$$E[X(s)X(t)] = P(X(s) = X(t) = 1) = P(X(s) = 1, X(t) - X(s) = 0).$$

Now $X(t) - X(s) = 0$ only if there is an even number of jumps in $(s,t]$, i.e., $Y(t) - Y(s)$ is even. Using the independence between $X(s)$ and the Poisson distributed increment $Y(t) - Y(s)$, with expectation $\lambda(t - s)$, we get[2] for $\tau = t - s > 0$,

$$
\begin{aligned}
\mathsf{E}[X(s)X(t)] &= \mathsf{P}(X(s) = 1) \cdot \mathsf{P}(Y(t) - Y(s) \text{ is even}) \\
&= \frac{1}{2}\mathsf{P}(Y(t) - Y(s) \text{ is even}) = \frac{1}{2} \sum_{k=0,2,4,\ldots} e^{-\lambda\tau}(\lambda\tau)^k/k! \\
&= \frac{1}{4}e^{-\lambda\tau}\{e^{\lambda\tau} + e^{-\lambda\tau}\} = \frac{1}{4}(1 + e^{-2\lambda\tau}).
\end{aligned}
$$

For $0 < t < s$ we just replace $t - s$ with $s - t = -(t - s) = |t - s|$.

Conclusion: The random telegraph signal is weakly stationary with mean $m = \mathsf{E}[X(t)] = 1/2$, and exponentially decreasing covariance function

$$
r(\tau) = \mathsf{E}[X(s)X(s + \tau)] - m^2 = \frac{1}{4}e^{-2\lambda|\tau|}.
$$

In Example 2.9 we saw an example of a realization of a random telegraph signal with this covariance function. ▲

3.4 The covariance intensity function

3.4.1 Correlation intensity

In a Poisson process $X(t)$, the intensity λ is proportional to the expected number of events per time unit; Equation. (3.3). But λ can also be interpreted as the probability of an event occurring during a very small time interval of width $h > 0$, $\mathsf{P}(X(t + h) = X(t) + 1) \approx \lambda h$, or, more precisely,

$$
\lim_{h \downarrow 0} \frac{\mathsf{P}(X(t + h) = X(t) + 1)}{h} = \lambda.
$$

This local event probability is constant, in agreement with the stationarity of the stream of events.

Now, one could go a step further and ask for the probability that there will be an event in each of two small intervals, separated by a distance τ. In fact, for the Poisson process, increments over disjoint intervals are independent (and therefore also uncorrelated), and hence, for $\tau > 0$,

$$
\lim_{h \downarrow 0} \frac{\mathsf{P}(X(t + h) = X(t) + 1 \text{ and } X(t + \tau + h) = X(t + \tau) + 1)}{h^2} = \lambda^2,
$$

---

[2]Remember $e^x = 1 + x + x^2/2! + x^3/3! + \ldots$.

$$\lim_{h\downarrow0} \frac{\mathsf{E}[(X(t+h)-X(t))(X(t+\tau+h)-X(t+\tau))]}{h^2} = \lambda^2, \tag{3.7}$$

$$\lim_{h\downarrow0} \frac{\mathsf{C}[X(t+h)-X(t),X(t+\tau+h)-X(t+\tau)]}{h^2} = 0. \tag{3.8}$$

One could say that, in a Poisson process, the *correlation intensity* is zero; there is no correlation between events, whatsoever. We will now briefly discuss counting processes with dependent events, for which the covariance limit in (3.8) is not zero.

3.4.2 Counting processes with correlated increments

Consider a counting process $\{X(t), t \geq 0\}$ that satisfies the conditions in Definition 3.2, except (a), so its increments may be dependent. This means that events in disjoint intervals can influence each other. We then strengthen the statement (b) in the definition, by requiring that the *joint distribution* of increments over disjoint intervals does not change with a common shift of time.[3] We say that the events form a *stationary stream of events*. It can then be shown[4] that the limit

$$\lim_{h\downarrow0} \mathsf{P}(X(t+h)-X(t)=1)/h = \lambda$$

exists and is independent of t. The constant λ is called the intensity, as for the Poisson process; it may be infinite.

Definition 3.3. *The function, defined for $\tau \neq 0$,*

$$w(\tau) = \lim_{h\downarrow0} \frac{\mathsf{P}(X(t+h)=X(t)+1, X(t+\tau+h)=X(t+\tau)+1)}{h^2} - \lambda^2$$

$$= \lim_{h\downarrow0} \frac{\mathsf{C}[X(t+h)-X(t), X(t+\tau+h)-X(t+\tau)]}{h^2}$$

is called the covariance intensity function (when the limit exists). The normalized function $w(\tau)/\lambda^2 \geq -1$ is called the correlation intensity function.

The correlation intensity function is a simple descriptor of the dependence

---

[3] When condition (a) holds, this is an automatic consequence of (b).
[4] See, for example, [16, Sec. 3.8].

between events. The interpretation is that if there is an event at time t, then the conditional probability for an event also in the small interval $[t + \tau, t + \tau + h)$ is approximately $h\lambda(1 + w(\tau)/\lambda^2)$. A negative $w(\tau)$ decreases the chance that two events occur at distance τ; a positive value increases the chance.

3.5 Spatial Poisson process

From the Poisson counting process $X(t)$, one can easily reconstruct the locations of the events in the underlying stationary stream of events.[5] As we saw in Section 3.2.3, the event times for the occurring events are uniformly, independently distributed over the observation interval. This property can be generalized to the concept of a *point process*, where points are selected at random. The points can be located on the positive real line, as is the case for the standard Poisson process. But it can also be points on the whole real line (for an example, see further on in Section 5.5.2), or in the plane, as in the Santa Claus Example 1.11, or on a sphere, for example the earth surface.

> **Definition 3.4** (Spatial Poisson process). *A spatial Poisson process in the plane $\mathbb{R}^2$ is a random distribution of event points over the real plane, such that the number of points in disjoint regions is statistically independent, and the number of points in a region A has a Poisson distribution with mean $\lambda\, m(A)$, where $m(A)$ is the area of the region. The parameter λ is the intensity, the average number of events per area unit.*

The counting process that corresponds to a spatial Poisson process is a *set indexed process*, $X(A)$. The study of such *point processes* is one of the great achievements in modern probability theory; see [17] for a thorough treatment and [12] for a statistically oriented presentation.

3.6 Inhomogeneous Poisson process

The Poisson process we have used until now is statistically *homogeneous*, in the sense that the distribution of the number of events in a time interval/region depends only on the length/area and not on its location; the intensity λ is constant. A very useful generalization is the *inhomogeneous* Poisson process with varying intensity, $\lambda(t)$ in time or $\lambda(x)$ in space. Its characteristic

---

[5]In fact, one does not always distinguish between the counting process and the equivalent stream of events – the term Poisson process is applied to both.

property is that the number of events in an interval/region A is Poisson distributed with expectation equal to $\int_A \lambda(x)dx$, and events in different regions are independent if the regions don't overlap. One should be aware that in an inhomogeneous Poisson process in time, the interarrival times are no longer (necessarily) independent or exponential.

Thinning

Suppose that events in a homogeneous Poisson process with intensity λ are deleted independently of each other, and with probability $q(x)$ depending on location, and retained with probability $p(x) = 1 - q(x)$. Then the result will be an inhomogeneous Poisson process. This procedure is naturally called "thinning." The intensity in the thinned process will be $\lambda(x) = \lambda p(x)$.

Overdispersion

A consequence of the varying intensity in an inhomogeneous Poisson process is that mean and variance of the number of events in disjoint intervals of the same length are not constant. If an inhomogeneous Poisson process is observed over a time interval $[0, N]$ and $x_1, \ldots, x_N$ are the number of events in the successive unit length intervals, then x_k is an observation of a Poisson variable X_k with mean $m_k = \int_{k-1}^{k} \lambda(t)\,dt$. Obviously $E[\sum_1^N X_k[=\sum_1^N m_k = m$, say, so $\frac{1}{N}\sum x_k$ is an unbiased estimate of the average intensity $\overline{m} = m/N$ over the observation interval. Further, $V[\sum_1^N X_k] = m$, since the sum has a Poisson distribution with variance equal to the mean. However, the common variance estimator, $s^2 = \frac{1}{N-1}\sum_1^N (x_k - \overline{x})^2$, is not an unbiased estimate of the variance m, but instead it is on the average greater. This property, called *overdispersion*, can be used for testing a constant intensity in a Poisson type process.

3.7 Monte Carlo simulation of Poisson processes

3.7.1 *Homogeneous Poisson processes in $\mathbb{R}$ and $\mathbb{R}^n$*

Poisson process in $\mathbb{R}$

Monte Carlo simulation of event times τ_k in a homogeneous Poisson process in $\mathbb{R}$ with intensity λ can take any of two characterizations as a starting point:

Exponential waiting times: According to Theorem 3.1, the interarrival times T_k in a Poisson process are independent and exponentially distributed. Thus, the following scheme will generate the event times τ_k in a Poisson process with intensity λ over an interval $[0, T]$:

    1. $\tau = 0, k = 1$,

2. Take u uniformly in $[0,1]$,
3. $\tau = \tau - \log(u)/\lambda$,
4. If $\tau > T$, STOP,
5. $\tau_k = \tau, k = k+1$,
6. Go to step 2.

Uniform conditional distribution: According to Remark 3.1, the arrival times in a homogeneous Poisson process, observed over an interval $[0,T]$, are distributed as an ordered sample of $N(T)$ independent variables, uniformly distributed over the observation interval. The size of the sample, $N(T)$, has a Poisson distribution with mean λT. The following scheme will generate the event times τ_k in $[0,T]$ from a sample of a random number of variables:

1. Take $N \in Poisson(\lambda T)$,
2. Take independent $x_1, \ldots, x_N$ uniformly in $[0,T]$,
3. Take $(\tau_1, \ldots, \tau_N) = \text{sort}\,(x_1, \ldots, x_N)$.

Poisson process in $\mathbb{R}^n$

For a spatial Poisson process in the plane or any higher dimension only the uniform conditional method lends itself to easy generalization. Take $n = 2$ as an example of how to generate a Poisson process in a bounded region $A \subset \mathbb{R}^2$.

1. Find a rectangle $\overline{A} = [0,T_1] \times [0,T_2] \supseteq A$,
2. Take $N \in Poisson(\lambda T_1 T_2)$,
3. Take independent $x_1, \ldots, x_N$ uniformly in $[0,T_1]$ and $y_1, \ldots, y_N$ uniformly in $[0,T_2]$,
4. Take those $(x_1, y_1), \ldots, (x_N, y_N) \in A$ as the sample.

3.7.2   Inhomogeneous Poisson processes

A straightforward way to generate a sample from an inhomogeneous Poisson process in $\mathbb{R}$ or $\mathbb{R}^n$ is by *thinning* a sample from a homogeneous process. Suppose the variable intensity is bounded by a finite constant, $\lambda(x) \le \lambda_{\max}, x \in A$. Points in a homogeneous Poisson process with intensity $\lambda_{\max}$ can then be simulated, for example, by means of the uniform conditioning technique. For each point in the sample one then decides whether it should be kept or deleted; a point at $x \in A$ is deleted with probability $q(x) = 1 - \lambda(x)/\lambda_{\max}$.

For an inhomogeneous Poisson process in $\mathbb{R}$ one can consider a generalization of the uniform conditioning method. Suppose we want to simulate the process in an interval $A = [0,T]$. If $\Lambda(t) = \int_0^t \lambda(x)\,dx$ is the integrated intensity, then $F(x) = \Lambda(x)/\Lambda(T)$ is a cumulative distribution function (CDF). If it is simple to generate random variables from this distribution one can use the following scheme.

1. Take $N \in Poisson(\Lambda(T))$,
2. Take independent $x_1, \ldots, x_N$ according to $F(x) = \Lambda(x)/\Lambda(T)$,
3. Take $(\tau_1, \ldots, \tau_N) = $ sort $(x_1, \ldots, x_N)$.

Exercises

3:1. Let $\{X(t)\}$ be a Poisson process with intensity $\lambda > 0$. Since the mean of the process is $m_X(t) = \lambda t$ we can conclude that the Poisson process is not weakly stationary. Is the process $Y(t) = X(t) - \lambda t$ weakly stationary?

3:2. In a Poisson process one has observed four events in the interval $(0, 4]$. Compute the probability that two occurred in $(0, 1]$ and the remaining two in $(1, 2]$.

3:3. Fires in passenger trains are assumed to occur according to a Poisson process. During the period January 1, 1986 to December 31, 1994, there were 67 fires in passenger trains in Sweden. The total distance traveled by passenger trains was 544×10^6 km. Estimate the intensity of passenger train fires (per train km), and construct a two-sided confidence interval with approximate confidence level 95%. You may use the normal approximation.

3:4. A Geiger counter is used to measure radioactive decay. To estimate the activity in a radioactive sample one can do the following experiment. First, the natural background radiation is measured during a time period T_b and then, during a time period T_s, one measures the sample plus background.

Assuming both background and sample radiation are Poisson processes, with intensities λ_b and λ_s, respectively, explain how the sample activity can be estimated. Also, for a fixed total experiment time $T = T_b + T_s$, find the best choice of T_b and T_s that makes the variance of the estimate as small as possible. Finally, construct an approximate confidence interval for λ_s.

3:5. Let $\{X(t)\}$ be a Poisson process with intensity λ and denote by T_1 the time of the first event. Then, $P(T_1 \leq s) = 1 - \exp(-\lambda s)$. Show that, for $0 < s < t$,

$$P(T_1 \leq s \mid X(t) = 1) = s/t.$$

3:6. Let $\{X(t)\}$ be an inhomogeneous Poisson process with time dependent intensity $\lambda(t) = \gamma t$. Find, with T_1 as in the previous problem, the probabilities $P(T_1 \leq s)$ and $P(T_1 \leq s \mid X(t) = 1)$.

3:7. The risk for car accidents on a specific road varies with the time of the day; denote by $\lambda(t)$ the intensity with which there occurs an accident at a certain road at time t, $0 \le t \le 24$.

a) What factors can influence $\lambda(t)$, directly or indirectly?

b) Assume $\lambda(t) = 0.001(2 - \cos 2\pi t/24)$, and find the average number of accidents per day, and the probability of no accidents between six in the evening and six in the morning.

3:8. A muffin with raisins may be viewed as part of a three-dimensional Poisson process with raisins as the points. How many raisins does a muffin need to contain, on the average, in order to make the probability that a muffin contains no raisins at most 0.05? What objection can you find against the use of a Poisson process for raisins muffins?

3:9. Germ-grain models are popular as models for spatial structures, for example in ecology and material science. In a simple germ-grain model, points are distributed over a region according to a Poisson process with intensity λ. With the Poisson points as centers, one places discs (or spheres in $\mathbb{R}^3$) with random sizes, independently of each other. One could think of trees of random size located in a forest according to a Poisson process.

A simple question is then: what is the probability that a point chosen at random is covered by at least one disc? Find that probability if the radiuses of discs are independent and uniformly distributed between 0 and $a > 0$.

3:10. Prove that every continuous function $m(t)$ that satisfies

$$m(s+t) = m(s) + m(t)$$

is of the form $m(t) = kt$. Hint: Observe that $m(2t) = 2m(t)$, and generalize.

Chapter 4

Spectral representations

4.1 Introduction

It may be fair to claim that the spectrum is *the* main instrument in the theory and application of stationary stochastic processes. The spectral decomposition of the covariance function gives a unique characterization of its properties in terms of a Fourier transform. An analogue is the Fourier transform of data, which predates the covariance approach. The view taken in this book is that data are observations of a stationary process. Both transforms represent a decomposition of a function (covariance function or data series, respectively) into a sum of cosine functions. In this chapter, we deal mainly with the Fourier transform of the covariance function, called the spectral density, but we also illustrate the corresponding transform of data. A thorough discussion of Fourier analysis of data will be given in Chapter 9.

In Chapter 2, we presented many examples of random mechanisms that generate a process and its covariance function. One of these mechanisms is the superposition of simple functions with only a slight element of randomness, such as cosine functions with different periods. A sum of many such independent "waves" tends to have a rather "random" behavior. This idea can be formalized, and leads to the spectral representation both of the covariance function and, in a more advanced theory, of the process itself.

Every covariance function in discrete time can be represented as a Fourier integral of a non-negative, symmetric, and integrable density function, called the spectrum or the spectral density; in continuous time one has to add the requirement that the covariance function is continuous. The integral of the spectral density is equal to the variance of the process, and it has the important physical meaning that it describes how the variance (= "power") is distributed over elementary components with different frequencies. For example, "band-limited white noise" (studied in Example 5.2 below) is a stationary process where all frequencies in a finite frequency interval are represented with the same power, in analogy with white light, which "contains all frequencies."

Processes in continuous time are treated in Section 4.2, and two special cases are identified: "continuous spectrum," when the spectral density is a common function, often continuous, and "discrete spectrum," when the spectral density is a sum of "delta functions" and the process itself is a sum of random cosine functions. For processes in discrete time, $\{X_n, n \in \mathbb{Z}\}$, the covariance function has a spectral representation as a Fourier integral over the bounded interval, $(-1/2, 1/2]$. This case is treated in Section 4.3.

If a continuous time process is sampled at regular time-points, with sampling distance d, then the spectral density of the sampled process extends over $(-f_n, f_n]$, where $f_n = 1/(2d) = f_s/2$ is the *Nyquist frequency*, equal to the maximal frequency that can be identified in the spectral representation in the sampled signal. Sampling of a continuous process is treated in Section 4.4. In Section 4.5, we have collected some special remarks, in particular about the precise relation between the covariance function and the spectrum.

4.2 Spectrum in continuous time

Every continuous covariance function for a stationary process has a Fourier transform (or, rather an inverse Fourier transform), called the spectral density.

4.2.1 Definition and general properties

Theorem 4.1. *If the covariance function $r(\tau)$ of a stationary process $\{X(t), t \in \mathbb{R}\}$ is continuous, there exists a positive, symmetric, and integrable function[1] $R(f)$ such that*

$$r(\tau) = \int_{-\infty}^{\infty} e^{i2\pi f\tau} R(f)\,df. \qquad (4.1)$$

The expression (4.1) is called the spectral representation of $r(\tau)$, and $R(f)$ is the spectral density of the covariance function $r(\tau)$ of the process $\{X(t)\}$.

The integral of the spectral density $R(f)$ is equal to the variance of the

---

[1]The function $R(f)$ may contain delta functions; properties of the delta function are described in Appendix B. In this chapter we use the following property of the delta function $\delta_{f_0}(f)$ located at f_0: $\int g(f)\delta_{f_0}(f)\,df = g(f_0)$ if the function $g(f)$ is continuous at $f = f_0$. For more on delta functions, see [41, Ch. II].

process,

$$V[X(t)] = r(0) = \int_{-\infty}^{\infty} R(f)\,df.$$

The (power) spectral density can be interpreted as a "variance decomposition"; it gives the "distribution" of the total variance (= power) over components with different frequencies. The integral

$$\int_{-b}^{-a} R(f)\,df + \int_{a}^{b} R(f)\,df = 2\int_{a}^{b} R(f)\,df \qquad (4.2)$$

is the variance contribution from the frequencies $0 \le a \le f \le b$.

Proof. The proof of the theorem requires advanced mathematics; see [33, Ch. 3]. First, one can observe that every covariance function is positive definite, which means that for every n, real numbers $a_1, \ldots, a_n$, and times $t_1, \ldots, t_n$,

$$\sum_{i,j=1}^{n} a_i a_j r(t_i - t_j) \ge 0. \qquad (4.3)$$

Then, one uses Bochner-Khinchin's theorem, which says that every continuous, positive definite function can be represented in the form (4.1). ∎

In Section 2.3.3, we left a loose end on what properties that are needed for a function to be a covariance function. The following theorem contains the converse of Theorem 4.1 and answers this question for continuous covariance functions.

Theorem 4.2. *If $R(f)$ is a non-negative, symmetric, and integrable function, possibly containing delta functions, then the function $r(\tau)$, defined by*

$$r(\tau) = \int_{-\infty}^{\infty} e^{i2\pi f\tau} R(f)\,df, \qquad (4.4)$$

is the covariance function of a stationary process.

Proof. One has to show that if $R(f)$ is a function with the stated properties, then $r(\tau)$, defined by (4.4), is positive definite and satisfies (4.3), and that then there is a stationary process that has $r(\tau)$ as its covariance function; in fact, one can find a Gaussian process with this property; for details, see [33, Ch. 3]. ∎

Remark 4.1. *Since the covariance function is a real function, the spectral representation can written in real form,*

$$r(\tau) = \int_{-\infty}^{\infty} \cos(2\pi f\tau) R(f)\,df + i \int_{-\infty}^{\infty} \sin(2\pi f\tau) R(f)\,df$$

$$= \int_{-\infty}^{\infty} \cos(2\pi f\tau) R(f)\,df = 2 \int_{0}^{\infty} \cos(2\pi f\tau) R(f)\,df, \quad (4.5)$$

since $e^{i2\pi f\tau} = \cos 2\pi f\tau + i\sin 2\pi f\tau$, and R is symmetric, i.e., $R(f) = R(-f)$. In practice, when a spectral density is given, it is usually wise to make sure if it is the one-sided $(2R(f))$ or the two-sided $(R(f))$ form that is presented.

Remark 4.2. *When t is a time parameter, f in $R(f)$ represents a* frequency, *with unit $[time\ unit]^{-1}$; for example, Hz if time is in seconds. One can also express the spectral representation by means of angular frequency, $\omega = 2\pi f$, as $r(\tau) = \int_{-\infty}^{\infty} e^{i\omega\tau} \widetilde{R}(\omega)\,d\omega$. The relations between the two spectral densities are simply*

$$\widetilde{R}(\omega) = \frac{1}{2\pi} R(\frac{\omega}{2\pi}), \qquad R(f) = 2\pi \widetilde{R}(2\pi f). \quad (4.6)$$

The name spectral representation is used for both forms.

It is also worth noting that a change of time scale in the process results in a corresponding change in the frequency scale. If $\{X(t), t \in \mathbb{R}\}$ has covariance function $r(\tau)$ and spectral density $R(f)$, then $X_c(t) = X(ct)$ for any constant c has covariance function $r_c(\tau) = r(c\tau)$ and spectral density $R_c(f) = c^{-1}R(f/c)$.

In the examples in Chapter 2, we studied two very different types of covariance functions: (a) $r(\tau) = \sigma^2 e^{-\alpha|\tau|}$ in Example 2.9, and (b) $r(\tau) = \sigma_0^2 + \sum_{k=1}^{n} \sigma_k^2 \cos(2\pi f_k\tau)$ in formula (2.7). The covariance function in (b) is a sum of periodic cosine functions, and it does not tend to 0 as $\tau \to \infty$, whereas the one in (a) does. This difference has its counterpart in the spectral density, and we shall consider the following two types of spectra:

a) **Continuous spectrum:** $R(f)$ is a continuous function, except for possible jump discontinuities; see Section 4.2.2;

b) **Discrete spectrum:** $R(f)$ is a sum of delta functions, i.e., there are constants b_k, f_k, $k = 0, \pm 1, \pm 2, \ldots$, such that $R(f) = \sum_k b_k \delta_{f_k}(f)$; see Section 4.2.4.

4.2.2 Continuous spectrum

There exists a simple condition on the covariance function that guarantees that the spectrum does not contain any delta functions. Under a somewhat stricter condition one can also compute the spectral density directly from the covariance function, as in the following theorem; see [33, Theorem 3.8].

Theorem 4.3. *(a) If the covariance function* $r(\tau)$ *satisfies* $\lim_{T\to\infty} T^{-1}\int_0^T r^2(t)\,\mathrm{d}t = 0$, *thus in particular if* $r(\tau) \to 0$ *as* $\tau \to \infty$, *then the spectrum does not contain any delta functions.*

(b) If the covariance function $r(\tau)$ *is absolutely integrable, i.e.,* $\int_{-\infty}^{\infty} |r(\tau)|\,\mathrm{d}\tau < \infty$, *the spectrum is continuous, and the spectral density is given by the Fourier inversion formula,*

$$R(f) = \int_{-\infty}^{\infty} e^{-i2\pi f\tau} r(\tau)\,\mathrm{d}\tau. \qquad (4.7)$$

Thus, the covariance function and the spectral density is a Fourier transform pair, $R = \mathscr{F}(r)$, $r = \mathscr{F}^{-1}(R)$.

(c) A continuous function $r(\tau)$ *with* $\int_{-\infty}^{\infty} |r(\tau)|\,\mathrm{d}\tau < \infty$ *is a covariance function if its Fourier transform* $R(f) = \int_{-\infty}^{\infty} e^{-i2\pi f\tau} r(\tau)\,\mathrm{d}\tau$ *is symmetric, non-negative, and integrable.*

Appendix D lists some useful Fourier transform pairs.

4.2.3 Some examples

Example 4.1 ("Low-frequency white noise"). The sinc-function, $\mathrm{sinc}\,\tau = \frac{\sin \pi\tau}{\pi\tau}$, is a covariance function, since it is continuous (the reader should show that!), and since it can be obtained from the spectral density

$$R(f) = \begin{cases} 1 & \text{for } -1/2 \le f \le 1/2, \\ 0 & \text{otherwise.} \end{cases}$$

We check this:

$$\int_{-\infty}^{\infty} e^{i2\pi f\tau} R(f)\,\mathrm{d}f = \int_{-1/2}^{1/2} e^{i2\pi f\tau}\,\mathrm{d}f = \frac{e^{i\pi\tau} - e^{-i\pi\tau}}{2i\pi\tau} = \frac{\sin \pi\tau}{\pi\tau}.$$

A process with this covariance function is an important special case of band-limited white noise, which will be treated in Example 4.3. ▲

Example 4.2 ("Ornstein-Uhlenbeck process"). The covariance function for the Ornstein-Uhlenbeck process in Example 2.9 is $r(\tau) = \sigma^2 e^{-\alpha|\tau|}$. We com-

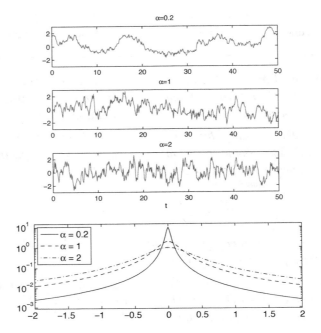

Figure 4.1 *Realizations of the Gaussian Ornstein-Uhlenbeck process with covariance function* $r(\tau) = e^{-\alpha|\tau|}$; α, *from top to bottom, is 0.2, 1, 2. The lower panel shows the corresponding covariance functions* $r(\tau)$.

pute its spectral density from (4.7):

$$R(f) = \int_{-\infty}^{\infty} \sigma^2 e^{-\alpha|\tau|} e^{-i2\pi f\tau}\, d\tau = \sigma^2 \left\{ \int_{-\infty}^{0} e^{(\alpha - i2\pi f)\tau}\, d\tau + \int_{0}^{\infty} e^{-(\alpha + i2\pi f)\tau}\, d\tau \right\}$$

$$= \sigma^2 \left\{ \frac{1}{\alpha - i2\pi f} + \frac{1}{\alpha + i2\pi f} \right\} = \sigma^2 \frac{2\alpha}{\alpha^2 + (2\pi f)^2}.$$

If α is large, then $R(f)$ is widely spread out, while if α is small, $R(f)$ is concentrated around the origin. In the former case, $r(\tau)$ decreases rapidly towards 0 as $|\tau| \to \infty$, while in the latter case, there is strong dependence even over long time spans.

Figure 4.1, upper panel, shows realizations of three Ornstein-Uhlenbeck processes, with α equal to 0.2, 1, and 2, respectively. The more rapid oscillations with increasing α are seen. The lower panel shows the corresponding spectral densities on logarithmic scale.   ▲

For easy reference we summarize the covariance/spectral density relation for the Ornstein-Uhlenbeck process (and for the random telegraph signal).

> **The Ornstein-Uhlenbeck process** has, with $\alpha > 0$,
>
> $$r_X(\tau) = \sigma^2 e^{-\alpha|\tau|}, \tag{4.8}$$
>
> $$R_X(f) = \sigma^2 \frac{2\alpha}{\alpha^2 + (2\pi f)^2}. \tag{4.9}$$

Example 4.3 ("Band-limited white noise"). A stationary process whose spectral density is constant over a finite interval (f_1, f_2),

$$R(f) = \begin{cases} c & \text{for } 0 < f_1 \leq |f| \leq f_2, \\ 0 & \text{otherwise,} \end{cases}$$

is called band-limited white noise. Such a process has variance $r(0) = \sigma^2 = \int_{-\infty}^{\infty} R(f)\,df = 2c(f_2 - f_1)$, and the covariance function

$$r(\tau) = \int_{-f_2}^{-f_1} e^{i2\pi f\tau} R(f)\,df + \int_{f_1}^{f_2} e^{i2\pi f\tau} R(f)\,df$$
$$= c \left[\frac{e^{i2\pi f\tau}}{i2\pi\tau} \right]_{f=-f_2}^{-f_1} + c \left[\frac{e^{i2\pi f\tau}}{i2\pi\tau} \right]_{f=f_1}^{f_2}$$
$$= \frac{c}{\pi} \left\{ \frac{\sin(2\pi f_2\tau)}{\tau} - \frac{\sin(2\pi f_1\tau)}{\tau} \right\}.$$

Inserting the c-value from the variance calculation, we get the covariance function (with $f_0 = (f_1 + f_2)/2$, $\Delta f = (f_2 - f_1)/2$),

$$r(\tau) = \frac{\sigma^2}{2\pi(f_2 - f_1)} \left\{ \frac{\sin(2\pi f_2\tau)}{\tau} - \frac{\sin(2\pi f_1\tau)}{\tau} \right\}$$
$$= \sigma^2 \cos(2\pi f_0\tau) \cdot \frac{\sin 2\pi\Delta_f\tau}{2\pi\Delta_f\tau}.$$

If the limits f_1 and f_2 are very close to each other, the process is called *narrow-banded noise*. Such a process apparently consists of oscillations with almost equal frequencies, and these tend to alternately amplify and extinguish each other. This gives the process a typical "narrow-banded" look, reminiscent of the sound of two slightly un-tuned music instruments.

If the lower frequency limit is $f_1 = 0$, the process is a low-frequency white noise with covariance function $r(\tau) = \sigma^2 \frac{\sin 2\pi f_2\tau}{2\pi f_2\tau}$. Despite its name, low-frequency white noise may contain very high frequencies, in absolute scale, if f_2 is large. All low-frequency white noise processes are time-scale changed versions of a standard process with $f_1 = 0, f_2 = 1/2$. If we let $f_2 \to \infty$, the covariance function $r(\tau) \to 0$ for $\tau \neq 0$, while $r(0) \to \infty$; see Section 6.4 on how to handle white noise with infinite upper frequency limit.

We have not mentioned anything about the distribution of the noise. Often, one assumes a normal distribution; i.e., the process is a Gaussian process, studied in detail in Chapter 5.   ▲

Example 4.4. This example illustrates the relation between the spectral density and the realizations of the process. Figure 4.2 shows realizations and spectral densities of three Gaussian processes $\{X_1(t)\}, \{X_2(t)\}, \{X_3(t)\}$, with the same variance 1. The processes are narrow-banded noise, low-frequency noise, and a Gauss-Markov Ornstein-Uhlenbeck process. The spectral densities are

$$R_1(f) = 4, \text{ for } 0.5 < |f| < 0.625, \quad \text{narrow-band noise in } (0.5, 0.625),$$
$$R_2(f) = \tfrac{1}{2}, \text{ for } |f| < 1, \quad \text{low-frequency noise in } (0, 1),$$
$$R_3(f) = \tfrac{2}{1+(2\pi f)^2}, \text{ for } -\infty < f < \infty, \quad \text{Ornstein-Uhlenbeck process}$$
$$\text{with } \alpha = 1.$$

The different degrees of noisiness are apparent, in particular for the Gauss-Markov process, which has no upper frequency limit.   ▲

Remark 4.3 ("Noise color"). *The analogy with colored light extends to more than just white noise/light, and different attributes are used depending on which side of the spectrum that dominates. In* red noise *low frequencies contribute most energy just as in red/infrared light. Its spectrum falls off rapidly in the high frequency end and is asymptotically proportional to* $1/f^2$, *as for the Ornstein-Uhlenbeck process. It is also called* brown noise.

Example 4.5 ("Ocean waves"). Ocean waves can be successfully modeled as stationary processes, at least if the wind has been reasonably constant over some time interval. The simplest stochastic wave model, the Gaussian model, assumes that elementary harmonic waves with different wavelengths are superimposed on each other, without interaction. The spectral density describes how the wave energy is distributed over the elementary waves with different wavelengths.

There are many types of standardized wave spectra designed for different parts of the ocean, depending on water depth, distance over which the wind can blow, and prevailing weather types. One such spectrum is the Jonswap spectrum, specially developed to describe North Sea waves; see [26].

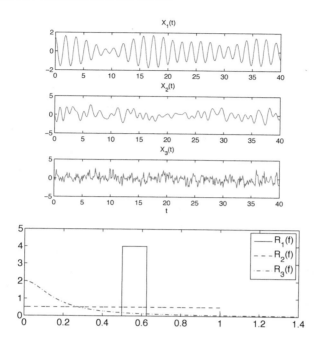

Figure 4.2 *Realizations of narrow-banded noise, low-frequency noise, and an Ornstein-Uhlenbeck process, with spectral densities R_1, R_2, and R_3, respectively.*

Figure 4.3 illustrates a spectrum for waves with average wave period 11 s and standard deviation 1.75 m.   ▲

Example 4.6 ("EEG"). In a diagnostic test, a patient with possible epilepsy diagnosis was subjected to periodic flashing light with different frequencies. The spectrum of the response EEG was calculated using one of the techniques

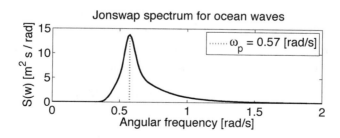

Figure 4.3 *Jonswap spectrum for ocean waves with peak angular frequency $\omega_0 = 0.57$ and average wave period $2\pi/\omega_0 = 11s$.*

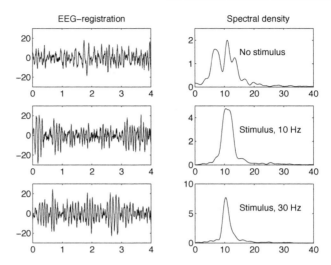

Figure 4.4 *EEG registrations (μV) (left) from a suspected epileptic patient with corresponding spectral densities (right); sampling frequency 256 Hz, measured over 4 seconds. Top: normal registration without stimulus; middle and bottom: light stimulus with frequency 10 Hz and 30 Hz.*

presented in Chapter 9. Figure 4.4 shows the results of the three phases in the experiment. The plot at the top shows the result of a session with open eyes without light stimulus. The two other plots illustrate two sessions with light stimulus with frequencies 10 Hz and 30 Hz, respectively. The estimated spectra show a clear response activity with a frequency of 10 Hz, increasing with increasing stimulus frequency. (We are grateful to Leif Sörnmo for permission to use this data.)   ▲

4.2.4 Discrete spectrum

The spectrum is discrete if the spectral density is a weighted sum of delta functions, with weights b_k, centered at discrete frequencies $\pm f_k$. We include the zero frequency, $f_0 = 0$, corresponding to a constant signal of random strength. The corresponding covariance function is

$$r(\tau) = \int e^{i2\pi f\tau} \sum_k b_k \delta_{f_k}(f)\,\mathrm{d}f = \sum_k b_k e^{i2\pi f_k \tau}. \qquad (4.10)$$

Since the spectral density is symmetric for a real-valued process, we can enumerate the frequencies so that $f_{-k} = -f_k$, $b_{-k} = b_k$, and write (4.10) using a real form, $r(\tau) = b_0 + \sum_{k=1}^{\infty} 2b_k \cos 2\pi f_k \tau$. In particular, $V[X(t)] = r(0) = \sum_{k=-\infty}^{\infty} b_k = b_0 + 2\sum_{k=1}^{\infty} b_k$.

A random harmonic oscillation

The process $X(t) = A_1 \cos(2\pi f_1 t + \phi_1)$, with random amplitude A_1 and phase ϕ_1 uniformly distributed in $[0, 2\pi]$, has covariance function $r(\tau) = \sigma_1^2 \cos 2\pi f_1 \tau$, as we saw in Section 2.4. It has all its power $r(0) = \sigma_1^2 = E[A_1^2]/2$, at the single frequency f_1, and the covariance function has a spectral representation with density

$$R(f) = \frac{\sigma_1^2}{2} \delta_{-f_1}(f) + \frac{\sigma_1^2}{2} \delta_{f_1}(f).$$

We verify this by means of Euler's formula, $\cos x = \frac{1}{2}(e^{-ix} + e^{ix})$:

$$r(\tau) = \frac{\sigma_1^2}{2}\left(e^{-i2\pi f_1 \tau} + e^{i2\pi f_1 \tau}\right) = \int_{-\infty}^{\infty} \left(\frac{\sigma_1^2}{2} \delta_{-f_1}(f) + \frac{\sigma_1^2}{2} \delta_{f_1}(f)\right) e^{i2\pi f \tau} \, df.$$

The spectral density is thus a sum of delta functions symmetrically located at $\pm f_1$, each with weight $\sigma_1^2/2$. Note, that we have split the variance $\sigma_1^2 = V[A_1 \cos(2\pi f_1 t + \phi_1)]$ equally between $-f_1$ and f_1.

Sums of random oscillations

A general process with discrete spectrum at $f_0 = 0, \pm f_k, k = 1, 2, \ldots, n$, can be realized as a sum of cosine functions with random phases and amplitudes,

$$X(t) = A_0 + \sum_{k=1}^{n} A_k \cos(2\pi f_k t + \phi_k), \tag{4.11}$$

where $E[A_0] = 0$, $V[A_0] = b_0$, $E[A_k^2] = 4b_k = 2\sigma_k^2$, for $k = 1, 2, \ldots$, with all variables independent. We showed in Section 2.4.2 that its covariance function is

$$r(\tau) = b_0 + \sum_{k=1}^{n} 2b_k \cos 2\pi f_k \tau, \tag{4.12}$$

which may be re-written

$$r(\tau) = b_0 + \sum_{k=1}^{n} b_k \left(e^{i2\pi f_k \tau} + e^{-i2\pi f_k \tau}\right) = \sum_{k=-n}^{n} b_k e^{i2\pi f_k \tau},$$

with the same convention as before ($b_{-k} = b_k$ and $f_{-k} = -f_k$).

   This is of type (4.10), and the process has discrete finite spectrum, with spectral density

$$R(f) = \sum_{k=-n}^{n} b_k \delta_{f_k}(f).$$

The periodic components in $X(t)$ contribute to the process's average power by the amount

$$V[A_0] = b_0 = \sigma_0^2, \quad V[A_k \cos(2\pi f_k t + \phi_k)] = 2b_k = \sigma_k^2,$$

respectively. Note, again, that the contributions have been split between the positive and negative frequencies.[2]

It is possible to let the sum (4.11) contain an infinite number of terms. We will investigate the meaning of an infinite sum of random variables in Example 6.5 in Chapter 6, but for now we just state that

$$X(t) = A_0 + \sum_{k=1}^{\infty} A_k \cos(2\pi f_k t + \phi_k) \tag{4.13}$$

means that

$$E\left[\left(X(t) - A_0 - \sum_{k=1}^{n} A_k \cos(2\pi f_k t + \phi_k)\right)^2\right] \to 0 \quad \text{as} \quad n \to \infty.$$

The constructive definition (4.13) of a stationary process can be generalized to integral form. The idea is to consider a sequence of processes of the form (4.13), where the power σ_k^2 contributed by a component with frequency f_k is successively spread out on a finer and finer net of frequencies. The result is an integral $X(t) = \int_{-\infty}^{\infty} e^{i2\pi f t} \, dZ(f)$, where $Z(f)$ is a complex valued random *spectral process* which defines infinitesimal independent amplitudes and uniformly phases to a continuum of frequencies; for details, see [24, 33].

4.3   Spectrum in discrete time

4.3.1   Definition

Let $\{X_t, t = 0, \pm 1, \pm 2, \ldots\}$ be a weakly stationary sequence. If it contains periodic components, then the length of one period must be at least 2, or equivalently, the frequency can be no more than $1/2$. Except from this upper bound, spectral theory is similar, but somewhat simpler, than for processes with continuous time.

---

[2]The introduction of negative frequencies is here made only to facilitate some calculations. If one is interested in stochastic processes in both time and space, for example random moving waves, there is another reason to separate positive and negative frequencies: the sign of the frequency determines the direction of the wave movement; for more on this, see [33, Ch. 3]. Negative frequencies also occur for complex-valued time series.

Theorem 4.4. *(a) For every covariance function $r(\tau)$ of a stationary sequence $\{X_t, \; t = 0, \; \pm 1, \pm 2, \dots\}$, there is a positive and integrable density function $R(f)$, defined on the half open interval $(-1/2, 1/2]$, and symmetric on the open interval $(-1/2, 1/2)$, such that*

$$r(\tau) = \int_{-1/2+0}^{1/2} e^{i2\pi f \tau} R(f) \, df.$$

(b) A function $r(\tau)$, $\tau = 0, \pm 1, \pm 2, \dots$, with $\sum_{-\infty}^{\infty} |r(\tau)| < \infty$, is a covariance function if its Fourier transform

$$R(f) = \sum_{-\infty}^{\infty} e^{-i2\pi f \tau} r(\tau) \qquad (4.14)$$

is symmetric, non-negative, and integrable.

The proof is similar to the proof of Theorem 4.1, except that Bochner-Khinchin's theorem is replaced by Herglotz' Lemma, its counterpart in discrete time. As in the continuous case, the function $R(f)$ is called the spectral density, and it may contain delta functions.

Example 4.7 ("White noise in discrete time"). If the X_t-variables are uncorrelated, then for integer τ,

$$r(\tau) = \begin{cases} \sigma^2, & \tau = 0, \\ 0, & \tau \neq 0. \end{cases}$$

The condition for the inversion formula is obviously satisfied, and the spectral density is

$$R(f) = \sigma^2, \quad \text{for} \quad -1/2 < f \leq 1/2.$$

Hence, the spectral density of a stationary sequence of uncorrelated variables is constant. Such a sequence is called white noise in discrete time, still in analogy with white light. Colored noise has non-constant spectral density. ▲

In analogy with continuous time, a stationary sequence with discrete spectrum may be written as a sum $X_t = \sum_k A_k \cos(2\pi f_k t + \phi_k)$, with the obvious difference that t is integer valued and $0 \leq f_k \leq 1/2$. The covariance function has a spectral representation with delta functions,

$$r(\tau) = \sum_k \frac{E(A_k^2)}{2} \cos 2\pi f_k \tau = \int_{-1/2+0}^{1/2} e^{i2\pi f \tau} \left\{ \sum_k \frac{E(A_k^2)}{4} \delta_{\pm f_k}(f) \right\} df.$$

A stationary sequence with discrete spectrum is a sum of sampled cosine functions with different amplitudes, frequencies, and phases.

4.3.2   Fourier transformation of data

In this book, we have adopted a probabilistic view on correlation and spectra, as properties of statistical distributions in a stochastic process model that generates data. One can also start from the data side, and use correlation and spectrum as means to summarize essential properties of the data; this is the approach in the classic book [4]. The aim of this section is to illuminate the relation between the statistical approach to stationary processes, where the spectrum is the Fourier transform of a covariance function, and the signal processing approach, starting with the experiment and the Fourier transformation of observed data.

The origin of stationary processes lies in data, but in the early days of practical Fourier analysis, the approach to "hidden periodicities" in a data series was clearly deterministic. The statistical process theory was developed later, when one started to realize the need for objective criteria to validate the findings; see the comments in the introduction to Chapter 7 about G.U. Yule and G.T. Walker, and the development of statistical time series analysis. In Chapter 9, on frequency analysis, we combine the statistical and data analytic approaches.

In the probabilistic approach to stationary processes, the covariance function is defined as an expectation, $r(\tau) = C[X_t, X_{t+\tau}] = E[(X_t - m)(X_{t+\tau} - m)]$. The spectral density $R(f)$ is *defined to satisfy the relations*

$$r(\tau) = \int_{-1/2}^{1/2} e^{i2\pi f\tau} R(f)\,\mathrm{d}f, \quad R(f) = \sum_{\tau=-\infty}^{\infty} e^{-i2\pi f\tau} r(\tau).$$

In the data analysis approach to Fourier analysis, one starts with a data series and calculates its Fourier transform. We shall now see what happens when the data are generated by a stochastic process with discrete spectrum.

Suppose that $\{X(t), t \in \mathbb{R}\}$ is a stationary process with discrete spectrum, concentrated at the equidistant frequencies $f_k = k/n$, $k = 0,\ldots,n/2-1$, between 0 and $1/2$. According to (4.11), the process can be explicity generated in continuous time as

$$X(t) = \sum_{k=0}^{n/2-1} A_k \cos(2\pi f_k t + \phi_k), \tag{4.15}$$

with random A_k, ϕ_k.

Assume now that we observe the process only at equidistant times $t = 0, 1, \ldots, n-1$. Then the observed values are

$$x_t = \sum_{k=0}^{n/2-1} a_k \cos(2\pi f_k t + \phi_k), \qquad (4.16)$$

where a_k are realizations of the random amplitudes and ϕ_k are realizations of the phases (note that for the phases we, as is customary, use the same symbol for the stochastic variable and for its realization).

The (inverse) Fourier transform of the data series is the *Discrete Fourier Transform*, DFT (see (9.20) in Chapter 9 for more details). Its value for $f_{k_0} = k_0/n$, $k_0 = 1, \ldots, n/2 - 1$, is equal to the

$$Z_n(k_0/n) = \sum_{t=0}^{n-1} x_t e^{-i2\pi(k_0/n)t} =$$

$$= \sum_{k=0}^{n/2-1} \frac{a_k}{2} \left\{ e^{i\phi_k} \sum_t e^{i2\pi(k-k_0)t/n} + e^{-i\phi_k} \sum_t e^{-i2\pi(k+k_0)t/n} \right\} = \frac{n}{2} a_{k_0} e^{i\phi_{k_0}}.$$

For $k_0 = 0$ one gets $Z(0) = \sum x_t = \frac{n}{2} a_0 \cos \phi_0$.

The normalized Fourier transform of the data series, $\frac{1}{n} Z_n(k/n)$, calculated at $f_{k_0} = k_0/n$, thus isolates the frequency specific amplitudes a_{k_0} and phases ϕ_{k_0}. Its squared absolute value is $\frac{1}{n} |Z_n(k_0/N)|^2 = a_{k_0}^2/4$. We observe that the normalized sum

$$I = \frac{\sum_{k=1}^{n/2-1} \frac{1}{n} |Z_n(k/n)|^2}{n/2 - 1} = \sum_{k=1}^{n/2-1} a_k^2/2 \qquad (4.17)$$

is equal to the average AC power per time unit of the signal, c.f., (2.8).

The expression $\frac{1}{n} |Z_n(k/n)|^2$ is called the *periodogram* of the data series, and it will be studied in detail in Chapter 9. With the amplitudes a_k as observations of the random variables A_k, we find the expectation of the averaged periodogram as

$$E[I] = \sum_k E[A_k^2]/2 = V[X(t)]. \qquad (4.18)$$

This relation helps us to regard the periodogram as a decomposition of the total variance of the process into contributions from discrete harmonic components. With a term borrowed from statistical methodology, we could say that we have performed an *analysis of variance*.

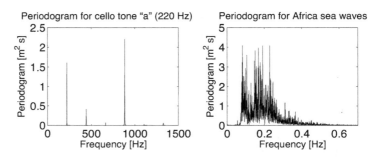

Figure 4.5 *Periodogram from discrete and continuous spectra. Left: a cello tone "a" (220Hz, including overtones); periodogram scaled by a factor 1000. Right: water waves at the coast of Africa.*

Example 4.8 ("Discrete and continuous spectrum"). This example illustrates the main difference between discrete and continuous spectrum.

Figure 4.5 contains periodograms from a cello tone "a" (220Hz), and from ocean water waves. The tone "a" consists of one fundamental tone plus a number of distinct overtones, where the one at 880 Hz is the strongest. It is reasonable to consider this as a process of the type (4.15).[3] The amplitudes are non-zero only for those frequencies that are present in the sum. This is not exactly true. If a frequency happens to fall between two values k/N and $(k+1)/N$, then two spikes will appear in the periodogram, each about half the real spike; more on this in Chapter 9. The periodogram for the water waves is quite different, and here it is reasonable to regard the underlying spectrum as continuous.

The frequency axes in the figure have been adjusted to match the physical frequency scales.   ▲

4.4   Sampling and the aliasing effect

Of course, there are many situations where data come as a discrete time sequence, for example, in an economic time series of daily, monthly, or yearly sales. In other cases, the data sequence has been obtained by *sampling* a continuous time process at discrete and equidistant time points.[4] It is then obvious that the sampling procedure can not catch rapid fluctuations between the sampling points, and that therefore the sampled sequence can have a quite

---

[3]In reality, the phases of the different overtones are not independent but related to each other in a way characteristic for the instrument.

[4]Non-equidistant sampling is possible, but makes statistical estimation more difficult; see [5].

different frequency profile than the continuous time process. Denote the sampling interval by d and the corresponding *sampling frequency* by $f_s = 1/d$, and define the *Nyquist frequency* $f_n = 1/(2d)$. The following theorem gives the exact relation between the spectra in discrete and continuous time.

Let $\{Y(t), t \in \mathbb{R}\}$ be a stationary process in continuous time with spectral density $R_Y(f)$ and covariance function $r_Y(\tau)$; thus $r_Y(0) = V[Y(t)] = \int_{-\infty}^{\infty} R(f) \, df$.

Let d be a sampling interval, observe $Y(t)$ at times $\ldots, -2d, -d, 0, d, 2d, \ldots$, and define

$$Z_t = Y(t), \ t = 0, \pm d, \pm 2d, \ldots.$$

Obviously, the sequence $\{Z_t, t = \ldots, -2d, -d, 0, d, 2d, \ldots\}$ has the same mean function and covariance function as the Y-process at the sample points, i.e., for $\tau = 0, \pm d, \pm 2d, \ldots$, i.e.,

$$r_Z(\tau) = C[Z_t, Z_{t+\tau}] = C[Y(t), Y(t+\tau)] = r_Y(\tau) = \int_{-\infty}^{\infty} e^{i2\pi f \tau} R_Y(f) \, df.$$

Thus, we have represented a covariance function in discrete time as a Fourier integral over the entire real line. One could ask how this agrees with the result of Section 4.3, where the spectrum for a stationary sequence extended only over the finite interval $(-1/2, 1/2]$. The answer is found in the following theorem.

Theorem 4.5. *The spectrum of the sampled process* $\{Z_t, t = 0, \pm d, \pm 2d, \ldots\}$ *can be concentrated in the interval* $-f_n < f \le f_n$, *where* $f_n = 1/(2d)$ *is the Nyquist frequency. The covariance function* $r_Z(\tau)$ *can be expressed as*

$$r_Z(\tau) = \int_{-f_n+0}^{f_n} e^{i2\pi f \tau} R_Z(f) \, df,$$

for $\tau = 0, \pm d, \pm 2d, \ldots$, *with a spectral density*

$$R_Z(f) = \sum_{k=-\infty}^{\infty} R_Y(f + k f_s), \ \text{for} \ -f_n < f \le f_n. \qquad (4.19)$$

Proof. Consider the spectral representation of $r_Z(\tau) = r_Y(\tau)$, when $\tau = nd$,

$$r_Z(\tau) = r_Y(nd) = \int_{-\infty}^{\infty} e^{i2\pi f n d} R_Y(f) \, df = \sum_{k=-\infty}^{\infty} \int_{(2k-1)f_n}^{(2k+1)f_n} e^{i2\pi f n d} R_Y(f) \, df$$

$$= \sum_{k=-\infty}^{\infty} \int_{-f_n}^{f_n} e^{i2\pi(f+kf_s)nd} R_Y(f+kf_s)\,\mathrm{d}f$$

$$= \int_{-f_n}^{f_n} \sum_{k=-\infty}^{\infty} e^{i2\pi(f+kf_s)nd} R_Y(f+kf_s)\,\mathrm{d}f.$$

Since $e^{i2\pi(f+kf_s)nd} = e^{i2\pi fnd}$, this is equal to

$$\int_{-f_n}^{f_n} e^{i2\pi fnd} \left\{ \sum_{k=-\infty}^{\infty} R_Y(f+kf_s) \right\}\mathrm{d}f,$$

and hence $r_Z(\tau)$, with $\tau = nd$, can be expressed as a Fourier transform of

$$R_Z(f) = \sum_{k=-\infty}^{\infty} R_Y(f+kf_s),$$

with f in the interval $(-f_n, f_n]$. ∎

Remark 4.4. *("The aliasing effect"). Note how the spectral density $R_Z(f)$ has been obtained by moving the spectrum contributions for all intervals with length $f_s = 1/d$ to the right and to the left of the central interval $(-f_n, f_n]$, and placing them over the central interval, adding them to $R_Z(f)$. All frequencies $f_0 + kf_s$ for $k = 1, 2, \ldots$ (to the right), and for $k = -1, -2, \ldots$ (to the left), are identified with the main frequency f_0. In the total integral, all frequencies $f_0 + kf_s$ have been* aliased *with f_0; see Figure 4.6.*

Note, also, that a covariance function $r_Z(\tau)$, which is defined only for discrete $\tau = 0, \pm d, \pm 2d, \ldots$, can be represented as a Fourier integral from $-\infty$ to $+\infty$ in many different ways. Of course, $r_Z(\tau) = \int e^{i2\pi f\tau} R_Z(f)\,\mathrm{d}f$, if we define $R_Z(f) = 0$ for $f \le -f_n$ and $f > f_n$. By adding $R_Y(f)$ in other ways than in (4.19), one can express $r_Z(\tau)$ (and also Z_t) by means of quite other frequencies, for example, $f_n < f < 3f_n$; cf. Figure 4.7. For obvious reasons, the aliasing effect is also called the folding *effect, since the spectrum is folded repeatedly around $\pm f_n$.*

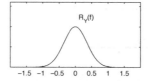 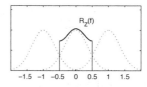

Figure 4.6 *Spectral density R_Z after sampling is generated by repeatedly folding the continuous time spectral density $R_Y(f)$ around the Nyquist frequency $\pm f_n$. In the figure $d = 1$ and hence $\pm f_n = 1/2$.*

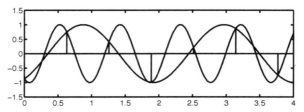

Figure 4.7 *It is not possible to differentiate between cosine functions with frequency difference k/d if one samples only at $t = 0, \pm d, \pm 2d, \ldots$.*

Normalized frequency

Instead of indexing the sampled observations Z_t in Theorem 4.5 by $t = 0, \pm d, \pm 2d, \ldots$, with the original time scale, one can use the sampling distance d as new time unit, and define

$$X_k = Z_{kd} = Y(kd), \quad k = 0, \pm 1, \pm 2, \ldots.$$

The covariance function is, of course, $r_X(k) = r_Z(kd) = r_Y(kd)$, and a change of variables, $v = f/f_s$, in the spectral representation yields the representation

$$r_X(\tau) = \int_{-f_n}^{f_n} e^{i2\pi f \tau d} R_Z(f) \, df = f_s \int_{-1/2}^{1/2} e^{i2\pi v \tau} R_Z(v/d) \, dv,$$

for $\tau = 0, \pm 1, \pm 2, \ldots$. The spectral density of the sequence $\{X_k\}$ is, therefore,

$$R_X(v) = f_s R_Z(v/d) = f_s \sum_{k=-\infty}^{\infty} R_Y\left(\frac{v+k}{d}\right), \quad -1/2 < v \leq 1/2.$$

The normalized frequency $v = f/f_s$ in this representation is dimensionless.

Under-sampling

The aliasing effect can be disturbing if the process contains powerful components with frequencies higher than the Nyquist frequency, $f_n = 1/(2d)$, which will be aliased with the components in the main frequency band. In practice, one often filters the signal through a low-pass filter before sampling. How this is done is described in Chapter 6.

If the process is *under-sampled*, i.e., the spectral density is not zero or almost zero, for frequencies above the Nyquist frequency, the observations can be seriously distorted as exemplified in Figure 4.8. The two top rows show spectral densities and resulting sequences when sampling with fast enough sampling rate. In the third row the sampling rate is slightly too low. The

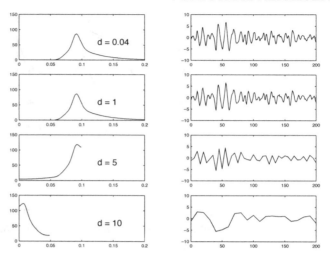

Figure 4.8 *Four examples of folded spectra $R_Z(f)$ (to the left) and sampled sequences (to the right) with sampling intervals $d = 0.04, d = 1, d = 5, d = 10$.*

bottom row gives a completely wrong impression where all power has been moved to a low-frequency band. The distortion is related to the well-known phenomenon in wild west movies, where the wheels of the prairie wagons seem to go backwards.

Example 4.9 ("Low-frequency white noise"). If low-frequency white noise $\{Y(t), t \in \mathbb{R}\}$ with spectral density and covariance function

$$R_Y(f) = \begin{cases} 1 & \text{for } |f| \le 1/2, \\ 0 & \text{otherwise,} \end{cases} \quad \text{and} \quad r_Y(\tau) = \frac{\sin \pi \tau}{\pi \tau},$$

is sampled with sampling distance $d = 1$, there will be no folding effect, the sampled sequence will have constant spectral density in $(-1/2, 1/2]$, and covariance function $r_Z(\tau) = r_Y(\tau) = 0$ for $\tau = \pm 1, \pm 2, \ldots$, $r_Z(0) = 1$, i.e., it will be white noise. With sampling distance $d = 1.5$, i.e., $1/(2d) = 1/3$, the spectral density is

$$R_Z(f) = \begin{cases} 2 & \text{for} & -1/3 < f \le -1/6, \\ 1 & \text{for} & -1/6 < f \le 1/6, \\ 2 & \text{for} & 1/6 < f \le 1/3, \\ 0 & \text{otherwise,} \end{cases}$$

with covariance

$$r_Z(\tau) = \frac{2(-1)^{\tau}}{3\tau\pi} \quad \text{for } \tau = \pm 1, \pm 2, \ldots. \qquad \blacktriangle$$

Example 4.10 ("White noise by sampling"). When a process $\{Y(t), t \in \mathbb{R}\}$ with

$$R_Y(f) = \begin{cases} 1 - |f| & \text{for } |f| \leq 1, \\ 0 & \text{otherwise,} \end{cases} \quad \text{and} \quad r_Y(\tau) = \left(\frac{\sin \pi \tau}{\pi \tau}\right)^2$$

is sampled with $d = 1$, then, since $\sin \pi = 0$, the spectral density and covariance function will be

$$R_Z(f) = \begin{cases} 1 & \text{for } -1/2 < f \leq 1/2, \\ 0 & \text{otherwise,} \end{cases} \quad \text{and} \quad r_Z(\tau) = 0 \quad \text{for } \tau = \pm 1, \pm 2, \ldots,$$

i.e., one gets white noise out of the sampling. ▲

Remark 4.5. *In signal processing it is common practice to regard the spectrum of a data sequence to be periodic, with period equal to* $2f_n = 1/d$, *when* d *is the length of the sampling interval. The inversion formula* (4.14)

$$R(f) = \sum_{-\infty}^{\infty} e^{-i2\pi f \tau} r(\tau)$$

(with $d = 1$*) can be used also for* f *outside the interval* $(-1/2, 1/2]$, *and it will produce a periodic function,* $R(f + k) = R(f)$. *Our definition of spectrum, both in continuous and in discrete time, requires that the integral of the spectrum is equal to the variance of the process, and this obviously excludes a periodic spectrum.*

4.5 A few more remarks and difficulties

4.5.1 The sampling theorem

In Section 4.4 we went from a continuous time process to discrete time by equidistant sampling. Here, we treat the inverse problem, how to construct a continuous time process and corresponding covariance function from a discrete time sequence.

The "random phase and amplitude" model expresses a stationary process as a sum of elementary cosine functions with random amplitudes and phases. If all amplitudes and phases are known, the process can be reconstructed completely from this discrete set of data. More surprising is, that also with a continuous spectrum, it is possible to reconstruct the entire process from a series of discrete observations. For a *band-limited* process, with spectrum restricted to an interval $[-f_0, f_0]$, this can be achieved by sampling the process at equidistant points with a sampling rate of at least $2f_0$, with sampling extending from minus infinity to plus infinity.

Suppose $r_0(kt_0)$, for $k \in \mathbb{Z}$, is the covariance function of a stationary sequence, defined at equally spaced times, $t = kt_0$, and that it has spectral density $R_0(f)$, for $-f_0 < f \leq f_0$, where $f_0 = 1/(2t_0)$. We assume that $R_0(f)$ does not contain any delta function at f_0. If the sequence in reality is sampled from a continuous time process, one could ask, what are the possible covariance functions for that process. One simple answer is that $r(\tau)$, defined by

$$r(\tau) = \int_{-f_0}^{f_0} e^{i2\pi f\tau} R_0(f) \, df, \text{ for } \tau \in \mathbb{R}, \tag{4.20}$$

is a covariance function in continuous time, and it coincides with $r_0(\tau)$ for $\tau = kt_0$, and it has the spectral density $R(f) = R_0(f)$ for $|f| \leq f_0$, and 0 otherwise.

In view of the aliasing theorem, Theorem 4.5, it is clear that there are many covariance functions that agree with $r_0(kt_0)$, but (4.20) is the one with the most concentrated spectrum. The following theorem is easily proved.

Theorem 4.6. *If $r_0(kt_0), k \in \mathbb{Z}$, is a covariance function in discrete time, then the continuous time covariance function (4.20) is equal to the interpolated*

$$r(\tau) = \sum_{k=-\infty}^{\infty} r_0(kt_0) \frac{\sin 2\pi f_0(\tau - kt_0)}{2\pi f_0(\tau - kt_0)}.$$

The simple covariance interpolation has a much deeper counterpart for the process itself, in the following famous theorem by Claude Shannon. It is formulated as a limit theorem for random variables, namely *convergence in quadratic mean*, $\mathsf{E}[(X_n - X)^2] \to 0$, defined in Appendix A.3, and also developed in Section 6.3.1.

Theorem 4.7 ("Shannon sampling theorem"). *If the stationary process $\{X(t), t \in \mathbb{R}\}$ is band-limited to $[-f_0, f_0]$, then it can be perfectly reconstructed from its values at discrete time points, spaced $t_0 = 1/(2f_0)$ apart. More precisely, for all t,*

$$X(t) = \lim_{N \to \infty} \sum_{k=-N}^{N} X(\alpha + kt_0) \cdot \frac{\sin 2\pi f_0(t - \alpha - kt_0)}{2\pi f_0(t - \alpha - kt_0)}, \tag{4.21}$$

where α is an arbitrary constant, and the limit is in quadratic mean.

The proof goes beyond the scope of this book; see [33, Section 5.2].

4.5.2 Fourier inversion

The spectral density $R(f)$ for a covariance function $r(\tau)$ can be a more or less "nice" function. It can be a continuous function, and it can be discontinuous and even unbounded, and it can contain delta functions. There is a balance between the smoothness of the spectral density, and the rate by which the covariance function $r(\tau)$ decays with increasing τ; if $r(\tau)$ is absolutely integrable, then the spectral density is continuous and can be calculated by means of the Fourier inversion formula, (4.7).

There is a more general theorem, with spectrum in integral form, that can be used for all covariance functions, with some careful interpretation. To formulate the theorem, we define a *spectral distribution function*, by

$$F_R(f) = \int_{-\infty}^{f} R(x)\,dx,$$

with the interpretation that if $R(f)$ contains a delta function at $f = f_0$ then $F_R(f_0)$ is set equal to the average of the left hand and right hand limits, $F_R(f_0) = (F_R(f_0 - 0) + F_R(f_0 + 0))/2$.

Theorem 4.8. *If $r(\tau)$ is a covariance function, then the following general inversion formulas hold,*
a) in continuous time:

$$F_R(f_2) - F_R(f_1) = \lim_{T \to \infty} \int_{-T}^{T} r(\tau) \frac{e^{-i2\pi f_2 \tau} - e^{-i2\pi f_1 \tau}}{-i2\pi\tau}\,d\tau,$$

b) in discrete time:

$$F_R(f_2) - F_R(f_1) = r(0)(f_2 - f_1) + \lim_{T \to \infty} \sum_{\substack{\tau = -T, \\ \tau \neq 0}}^{\tau = T} r(\tau) \frac{e^{-i2\pi f_2 \tau} - e^{-i2\pi f_1 \tau}}{-i2\pi\tau}.$$

These formulas can be used to compute a spectral density $R(f)$ by taking $f_2 = f + h$ and $f_1 = f - h$, and then let $h \to 0$. This directly gives any delta function that might be present. If the spectrum is continuous at $f = f_0$, with $R(f)$ a common function, one has to take the limit of the difference ratio, $\lim_{h \to 0}(F_R(f + h) - F_R(f - h))/(2h)$. The limit is equal to $R(f)$ if the density is continuous at f, and otherwise it is equal to the *Cauchy principal value* of the divergent integral (4.7), if the limit exists.

Example 4.11. We use the general inversion formula in Theorem 4.8, on the covariance function $r(\tau) = \frac{\sin 2\pi\tau}{2\pi\tau}$, which has spectral density $R(f) = \frac{1}{2}$ for $|f| \le 1$. For $h > 0$, we get, after some calculations,

$$\frac{F_R(f+h) - F_R(f-h)}{2h} = \frac{1}{2h} \lim_{T \to \infty} \int_{-T}^{T} \frac{\sin 2\pi\tau}{2\pi\tau} \frac{e^{-i2\pi(f+h)\tau} - e^{-i2\pi(f-h)\tau}}{-i2\pi\tau} \, d\tau$$

$$= \int_{-\infty}^{\infty} e^{-i2\pi f\tau} \frac{\sin 2\pi h\tau}{2\pi h\tau} \frac{\sin 2\pi\tau}{2\pi\tau} \, d\tau$$

$$= \begin{cases} \frac{1}{2} & \text{for } |f| < 1-h, \\ \frac{1}{4}(1 + (1-|f|)/h) & \text{for } 1-h < |f| < 1+h, \\ 0 & \text{for } |f| > 1+h. \end{cases}$$

The limit as $h \to 0$ is $1/2$, $1/4$, and 0, respectively, i.e., we get the spectral density as

$$R(f) = \begin{cases} 1/2 & \text{for } |f| < 1, \\ 1/4 & \text{for } |f| = 1, \\ 0 & \text{for } |f| > 1. \end{cases}$$

This is almost what we expected. The only difference is the averaged values at the discontinuity points. It is worth noting that if we use the Fourier inversion formula (4.7) directly on this covariance function, we get as Cauchy principal value,

$$\lim_{T \to \infty} \int_{-T}^{T} e^{-i2\pi\tau} r(\tau) \, d\tau = \lim_{T \to \infty} \int_{-T}^{T} \frac{\sin 4\pi\tau}{4\pi\tau} \, d\tau = 1/4,$$

for $f = 1$.   ▲

Remark 4.6 ("The aliasing effect"). *The fact that the spectral density $R(f)$ is not uniquely defined when the spectrum is continuous can cause some confusion when sampling a process in continuous time. The folding formula (4.19) holds if $R_Y(f)$ is continuous in all points $f + k/d$, but if not, one has to use the Cauchy principal value. For example, when sampling the low-frequency white noise in Example 4.11, with a sampling interval $d = 1$, formula (4.19) will give the bizarre answer*

$$R_Z(f) = \begin{cases} 1/2 & \text{for } -1 < f < 1, \\ 1 & \text{for } f = 1, \end{cases}$$

if one uses $R_Y(\pm 1) = 1/2$, while one gets the more reasonable value $R_Z(1) =$

1/2 if one uses $R_Y(\pm 1) = 1/4$. This slight indefiniteness in the spectral density has no practical consequences if the spectrum is continuous, since it is the integral of the spectral density that defines the variance contribution. If the spectrum is discrete it is uniquely determined.

4.5.3 Spectral representation of the second-moment function $b(\tau)$

The second-moment function $b(\tau) = E[X(t + \tau)X(t)]$ has a spectral representation in terms of a positive, symmetric, and integrable density function $B(f)$, such that

$$b(\tau) = \int_{-\infty}^{\infty} e^{i2\pi f\tau} B(f)\,df.$$

Since $b(\tau) = r(\tau) + m^2$, $(m = E[X(t)])$, there exists a simple relation between the spectral densities $R(f)$ for $r(\tau)$ and $B(f)$ for $b(\tau)$. With $B(f) = R(f) + m^2\delta_0(f)$, we have

$$\int_{-\infty}^{\infty} e^{i2\pi f\tau} B(f)\,df = \int_{-\infty}^{\infty} e^{i2\pi f\tau} R(f)\,df + m^2 \int_{-\infty}^{\infty} e^{i2\pi f\tau} \delta_0(f)\,df$$
$$= r(\tau) + m^2.$$

The only difference is thus the presence of the delta function $m^2\delta_0(f)$ in $B(f)$.

Example 4.12. Let $\{Y(t)\}$ be a stationary process with expectation 0 and with continuous spectrum with density $R_Y(f)$. Further, let Z be a random variable, independent of the Y-process, with mean 0 and variance σ_Z^2. Finally, define the process $X(t) = m + Z + Y(t)$, with expectation m, covariance function $r_X(\tau) = \sigma_Z^2 + r_Y(\tau)$, and second-moment function $b_X(\tau) = m^2 + \sigma_Z^2 + r_Y(\tau)$. The spectral density for the covariance function $r_X(\tau)$ is

$$R_X(f) = \sigma_Z^2\delta_0(f) + R_Y(f),$$

while the second-moment function $b_X(\tau)$ has spectral density

$$B_X(f) = m^2\delta_0(f) + \sigma_Z^2\delta_0(f) + R_Y(f) = (m^2 + \sigma_Z^2)\delta_0(f) + R_Y(f).$$

Both spectra contain a delta function at frequency 0, but with different weights, σ_Z^2 for $R_X(f)$, and $(m^2 + \sigma_Z^2)$ for $B_X(f)$. One would like to interpret the zero-frequency weight as the "average power at frequency 0." If we only know the spectrum $B_X(f)$ and the weight $m^2 + \sigma_Z^2$ we cannot differentiate between m^2, which comes from the deterministic expectation m, and σ_Z^2, which is the result of the random shift in average. This deficiency of the second-moment function is the main reason for our use of the covariance function. ▲

4.5.4   *Beyond spectrum and covariance function*

The machinery introduced so far in this book – weak stationarity, covariance function, and its Fourier transform, the spectrum – has obvious mathematical and statistical merits. The mathematical properties of the covariance function/spectral density pair are well understood and the representation of a process as a sum of non-interacting harmonic functions is easy to grasp and analyze. Only the second order moments are needed to define the model, and, in case the process is Gaussian, all properties can be derived from the covariance function. In particular, the Gaussian model allows the calculation of the probability of large deviations from the average value from the covariance function. Furthermore, as we will see in Chapter 6, spectral relations are easy to handle when a stochastic process is passed through a linear filter,

However, in many applications one observes occasional greater variability than expected from the covariance function and the normal distribution. Examples of such behavior are found in economics, structural and fluid mechanics, control systems, optics, and many other areas. In the stock market, calm periods with small variability in stock prices can be interrupted by periods of large fluctuations, making the distributions *heavy-tailed*. The truck in Example 1.6 can operate on a smooth road and suddenly enter a road section of bad quality. As a third example one can take a resonant system, like a pendulum, where a small regular variation in the length of the pendulum can cause a large response. The causes of the irregularities may be stochastic, and if the choice of origin in the time scale is irrelevant the statistical model may still be (strictly) stationary despite its seemingly non-stationary behavior.

There is no statistical theory that covers all such irregularities, and each application area has developed its own models and techniques. Many of the models contain some form of variation in the basic parameters, either caused by an external source or by the process itself. In physics and mechanics one uses the terms *parametric excitation* or *parametric instability*; a physical deformation of a mechanical structure may, for example, lead to a completely different system equation. In statistics a common term is *non-linear time series*. In Chapter 7 we describe a statistical model used in economics, the *GARCH* model.

4.6   Monte Carlo simulation from spectrum

See Section 5.6 for how to simulate a Gaussian process from the spectrum.

Exercises

4:1. Suppose that the stationary process $\{X(t), t \in \mathbb{R}\}$ has the covariance function $r_X(\tau) = e^{-\alpha|\tau|} \cos 2\pi f_0 \tau$, $\tau \in \mathbb{R}$. Find its spectral density.

4:2. Determine the corresponding spectral densities to the following covariance functions of stationary processes in continuous time:

$$(a)\ r_X(\tau) = e^{-\alpha \tau^2}, \qquad (b)\ r_X(\tau) = \frac{1}{1+\tau^2}.$$

4:3. A stationary process has the covariance function $r(\tau) = \alpha_0^2 + \frac{\alpha_1^2}{2} \cos 2\pi f_0 \tau$ in continuous time. Find the spectral density.

4:4. Find the covariance functions of processes with spectral densities
$$R_1(f) = \frac{A}{1+(2\pi f)^2}, \quad R_2(f) = \frac{A}{4+(2\pi f)^2}, \quad R_3(f) = \frac{A}{(1+(2\pi f)^2)(4+(2\pi f)^2)}.$$

4:5. Figure 4.9 shows four spectral densities and four covariance functions. Match up the corresponding spectral densities and covariance functions.

4:6. Figure 4.10 shows realizations of two stationary processes, $U(t)$ and $V(t)$, with the following spectral densities:

$$R_U(f) = \frac{200}{100^2 + (2\pi f)^2}, \quad f \in \mathbb{R},$$

$$R_V(f) = \begin{cases} 1, & 2 < |f| < 2.5, \\ 0, & \text{otherwise.} \end{cases}$$

(a) Find $C[U(t), U(t+\tau)]$ and $C[V(t), V(t+\tau)]$.
(b) Which realization belongs to which process?

4:7. Figure 4.11 shows three realizations, three covariance functions, and three spectral densities of three different stationary processes. Determine which figures that are realizations, covariance functions, and spectral densities. Also state how they are connected.

4:8. Find the values for the constant B such that

$$R(f) = \frac{1}{1+f^2} + \frac{B}{4+f^2}$$

is a spectral density in continuous time.

4:9. Figure 4.12 shows realizations, covariance functions, and spectral densities of two processes. One of the processes has continuous time and one has discrete time; both realizations have been smoothed to a continuous curve.

(a) For each plot, A–F, determine whether it depicts a realization, a covariance function, or a spectral density. Justify your answer!

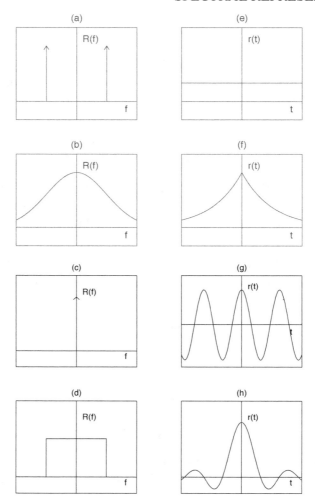

Figure 4.9 *Spectral densities and covariance functions to Exercise 4:5.*

(b) Which realization belongs to which covariance function and which spectral density?

(c) Suppose now that we would like to estimate the expected mean value of the processes with the average of only two samples at a time distance h. That is, we estimate $\mathsf{E}[X(t)] = m_X$ with $\widehat{m}_X = \frac{1}{2}(X(t) + X(t+h))$. Determine, for each process, an optimal value of h, i.e., the h giving the minimum variance of the estimator.

4:10. Suppose that $\{X(t), t \in \mathbb{R}\}$, t in seconds, is a process of interest with

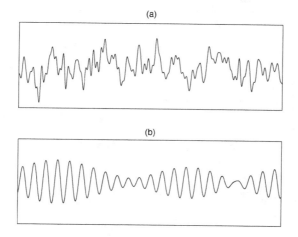

Figure 4.10 *Realizations to Exercise 4:6.*

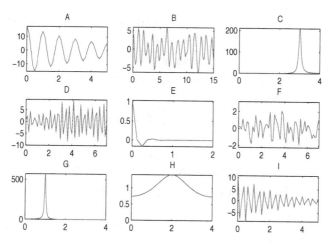

Figure 4.11 *Realizations, covariances, and spectra to Exercise 4:7.*

the following spectral density:

$$R_X(f) = \begin{cases} \cos^2 \frac{\pi f}{2} & \text{for } -1 \leq f \leq 1, \\ 0 & \text{otherwise.} \end{cases}$$

(a) Draw a graph of the spectral density of the process.
(b) In order to study the process, we need to sample it. How often should it be sampled if we want to avoid aliasing?

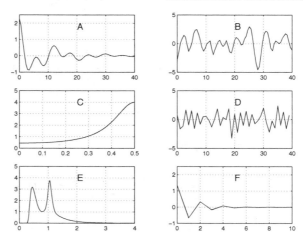

Figure 4.12 *Realizations, covariances, and spectra of processes in Exercise 4:9.*

(c) Compute and plot the spectral density of the sampled process, with (1) sampling once every second, (2) with five times per second?

4:11. The stationary process $\{X(t),\ t \in \mathbb{R}\}$ has a spectral density which we assume is constant $R_X(f) = C$ in the interval $10 \leq |f| \leq 20$ Hz and zero for all other values.

(a) Which is the smallest sampling frequency that can be chosen if one wants to avoid aliasing?

(b) Sometimes it is easy to forget that in many measurements there might be a strong disturbance at the frequency 50(60) Hz from the electricity current. Suppose that this disturbance process can be described by

$$N(t) = \cos(2\pi 50 t + \phi), \quad -\infty < t < \infty,$$

where the random variable ϕ is uniformly distributed in the interval $0 \leq \phi < 2\pi$. Our measurement signal consists of $Y(t) = X(t) + N(t)$ and we choose our sampling frequency to $1/d = 30$ Hz. What frequency interval is covered by the process $\{X(t)\}$ and the disturbance $\{N(t)\}$, respectively, after the sampling?

Chapter 5

Gaussian processes

5.1 Introduction

In a Gaussian process, all process values have normal distributions, and all linear operations, like summation, differentiation and integration, produce normally distributed random variables. Also, the derivative of a Gaussian process is a new Gaussian process.

A Gaussian process is completely determined by its mean value function and covariance function. Therefore, it is often the first choice when one seeks a model for a stochastic process, since estimation of these functions is a relatively simple task, as it involves only individual data and pairs of data. Estimation of higher moments, or more than bivariate data, is much more complicated or sensitive. Another reason for its popularity is that many phenomena in nature actually are Gaussian or almost Gaussian, or can be transformed to Gaussian by a simple transformation.

In Section 5.2, we define the Gaussian process by means of its mean value and covariance function, and discuss by examples the relation with the spectrum. We also describe a method to simulate a Gaussian process with a general spectrum, by approximation with a sum of cosines with random amplitudes and phases. Section 5.3 deals with the important Wiener process. (This process is often instead called a Brownian motion.) The envelope of a Gaussian process is briefly discussed in Section 5.4. It will studied in more detail in Section 8.3.

In Section 5.5, we present some generalizations, including the Lévy process, much used in financial statistics, and the shot noise process, used to describe fluctuations in electronic and optical devices.

Finally, in Section 5.6, we describe how one can simulate a sample from a stationary Gaussian process from a specified spectral density.

5.2 Gaussian processes

A Gaussian process has one main feature, that also can be used as definition: every real-valued linear operation applied to the process produces a normally distributed random variable. This sweeping statement is a consequence of the following formal definition.

Definition 5.1. *A stochastic process $\{X(t), t \in T\}$ is a Gaussian process if, for every n and all $t_1, \ldots, t_n \in T$, the vector $(X(t_1), \ldots, X(t_n))$ has an n-dimensional normal distribution.*

By Definition A.1 in Appendix A, an equivalent definition is that a process is Gaussian if every linear combination $a_1 X(t_1) + \cdots + a_n X(t_n)$ of process values has a normal distribution.

The definition implies that each single value $X(t)$, for fixed t, has a normal distribution, with mean and variance determined by the mean value and covariance functions:

$$X(t) \sim N(m(t), r(t,t)).$$

An n-dimensional normal distribution is determined by expectations and covariances. Therefore all n-dimensional distributions in a Gaussian process are uniquely determined by the mean and covariance functions:

$$
\begin{aligned}
m_k &= \mathsf{E}[X(t_k)] = m(t_k), \\
\sigma_{jk} &= \mathsf{C}[X(t_j), X(t_k)] = r(t_j, t_k).
\end{aligned}
$$

We can also draw the very important conclusion that, if a mean value function $m(t)$ and a covariance function $r(s,t)$ are given, a priori, then there always exists a Gaussian process that has these as mean and covariance functions; this is a consequence of Kolmogorov's existence theorem, Theorem C.1 in Appendix C.

As hinted in the introduction, one reason for the popularity of the Gaussian process is that linear operations, which are the topic of Chapter 6, produce normally distributed random quantities.

Summation, differentiation, and integration of Gaussian processes will, if the operation is possible, always generate a normal random variable or a new Gaussian process.

This is a simple consequence of the definition. Thus, if $\{X(t), t \in T\}$ is a Gaussian process, then, for example,

$$
\begin{aligned}
Y_1 &= X(t+h) + X(t), \\
Y_2 &= (X(t+h) - X(t))/h, \\
Y_3 &= (X(t_1) + \cdots + X(t_n))/n,
\end{aligned}
$$

all have normal distributions. Furthermore, (Y_1, Y_2, Y_3) has a three-dimensional normal distribution. Even limits of linear combinations are normally distributed, when they exist, for example,

$$
\begin{aligned}
X'(t) &= \lim_{h \to 0} \frac{X(t+h) - X(t)}{h}, \\
\int_0^1 X(t)\, dt &= \lim_{n \to \infty} \frac{1}{n} \sum_{k=1}^{n} X(k/n);
\end{aligned}
$$

see Section 6.3. (That limits of normal variables are also normally distributed can be shown by means of characteristic functions.)

5.2.1 Stationary Gaussian processes

A weakly stationary Gaussian process is also strictly stationary. In fact, if $\{X(t), t \in T\}$ is weakly stationary, then for arbitrary n and time points $\tau, t_1, \ldots, t_n$, weak stationarity means that the vectors $(X(t_1 + \tau), \ldots, X(t_n + \tau))$ and $(X(t_1), \ldots, X(t_n))$ have the same mean and the same covariances. Since a normal distribution is determined by its means and covariances, it follows that $(X(t_1 + \tau), \ldots, X(t_n + \tau))$ and $(X(t_1), \ldots, X(t_n))$ have the same distribution. Thus $\{X(t)\}$ is strictly stationary.

If $\{X(t)\}$ is a stationary Gaussian process with mean m and covariance function $r(\tau)$, all $X(t)$ have the same marginal distribution, the normal distribution, $N(m, r(0))$, and the probability density function is

$$
f_{X(t)}(x) = \frac{1}{\sqrt{2\pi r(0)}} \exp\left\{ -\frac{(x-m)^2}{2r(0)} \right\}.
$$

The two-dimensional density of $X(t_1)$ and $X(t_2)$, at distance $\tau = t_1 - t_2$ from each other, is expressed in terms of the variance $r(0)$ and the correlation coefficient $\rho = \rho(\tau) = r(\tau)/r(0)$,

$$
f_{X(t_1), X(t_2)}(x_1, x_2) = \frac{1}{c} \exp\{ -Q(x_1, x_2)/2 \},
$$

with $c = 2\pi r(0)\sqrt{1 - \rho^2}$, and

$$Q(x_1, x_2) = \frac{1}{r(0)(1 - \rho^2)} \left\{ (x_1 - m)^2 - 2\rho(x_1 - m)(x_2 - m) + (x_2 - m)^2 \right\}.$$

Example 5.1 ("Ornstein-Uhlenbeck process"). The function $r(\tau) = \sigma^2 e^{-\alpha|\tau|}$ is the well-known covariance function for the random telegraph signal from Example 3.7, which is not a Gaussian process, and the Ornstein-Uhlenbeck process, which is a Gauss-Markov process, being both Gaussian and Markovian. Figure 2.10(b) shows a realization of the Ornstein-Uhlenbeck process.   ▲

Example 5.2 ("Low-frequency and band-limited Gaussian noise"). Figure 5.1 shows realizations and covariance functions of two Gaussian processes, low-frequency and band-limited white noise, generated by linear filtering of genuine Gaussian electronic noise.[1] The covariance functions are

$$r(\tau) = \frac{\sin 2\pi\tau}{2\pi\tau},$$
$$r(\tau) = \frac{1}{2\pi(f_2 - f_1)} \left\{ \frac{\sin 2\pi f_2 \tau}{\tau} - \frac{\sin 2\pi f_1 \tau}{\tau} \right\},$$

respectively. Both processes are normalized to have $r(0) = 1$, and the constants are $f_1 = 1.8$, $f_2 = 2.2$, making the second process quite narrow-banded.

The band-limited realizations in Figure 5.1 deserve a comment on the meaning of "stationarity." That a stochastic process is stationary means that the random variation from realization to realization has the same distribution, regardless of when in time the observations are taken; i.e., $X(t_1)$ and $X(t_2)$ have the same distribution. This does not contradict the uneven fluctuations seen in the figure. A realization can stay almost constant for a while, and soon thereafter fluctuate vigorously. "Constant variance" means constant in time, between realizations. The local increase in variability seen in the figure is the result of random interference between frequency components, and it can be seen anywhere in time. The time origin has no special role.

The lesson to be learned from this example is that one should be very careful when comparing different short sections of an observed time series. They could be quite different just by chance, without any change in the generating mechanism or in the statistical properties.   ▲

---

[1]The upper diagrams in Figure 5.1 require a comment. As distinct from most pretty contemporary figures in this book they are real photos of physical electrical signals, generated by noise from a physical noise generator, filtered through an analogue electrical filter, displayed on an oscilloscope, and photographed with a Polaroid camera. They were part of the first version (1973) of the present work.

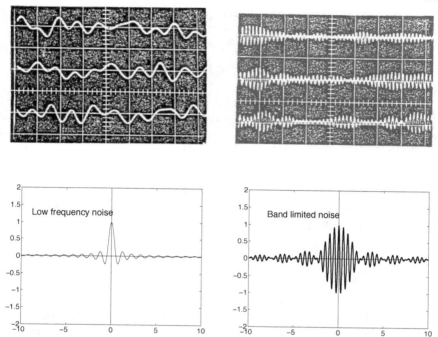

Figure 5.1 *Realizations and covariance functions for Gaussian low frequency and band-limited noise.*

Financial time series typically exhibit both quiet periods with only small variations, and periods with large variations (or, in financial language, large "volatility"), just as realizations of band-limited Gaussian noise. However, band-limited Gaussian noise is still too regular to be a useful model for financial time series, and one instead has to resort to non-linear time series models. The most important of these, the GARCH model, is briefly discussed in Section 7.5.

5.2.2 Gaussian process with discrete spectrum

We shall now see how the simple "random phase and amplitude" process $X(t) = A\cos(2\pi f t + \phi)$ can be constructed as a Gaussian process. Let X and Y be independent normal random variables, $N(0, \sigma^2)$. Then the process $\{X(t)\}$, defined by

$$X(t) = X\cos 2\pi f t + Y\sin 2\pi f t, \tag{5.1}$$

is a Gaussian process. It has mean 0 and covariance function $r(\tau)$ equal to

$$\mathsf{E}\big[\big(X\cos(2\pi fs)+Y\sin(2\pi fs)\big)\big(X\cos(2\pi f(s+\tau))+Y\sin(2\pi f(s+\tau))\big)\big]$$
$$= \sigma^2\big(\cos(2\pi fs)\cos(2\pi f(s+\tau))+\sin(2\pi fs)\sin(2\pi f(s+\tau))\big)$$
$$= \sigma^2\cos 2\pi f\tau.$$

We have already encountered this covariance function several times before. By re-arranging the Definition (5.1), we get the process into its standard form,

$$X(t) = A\{\cos\phi\,\cos 2\pi ft - \sin\phi\,\sin 2\pi ft\} = A\cos(2\pi ft + \phi), \qquad (5.2)$$

where the amplitude A and phase $\phi \in [0, 2\pi)$ are defined by $A = \sqrt{X^2 + Y^2}$, $\cos\phi = X/A$, $\sin\phi = -Y/A$. For normal X and Y, it is easy to show that A and ϕ are independent with densities,

$$f_A(u) = \tfrac{u}{\sigma^2}\exp\left(-\tfrac{1}{2}u^2/\sigma^2\right), \quad \text{for } u \geq 0, \qquad \text{“Rayleigh distribution,”}$$
$$f_\phi(v) = \tfrac{1}{2\pi}, \qquad\qquad\qquad \text{for } 0 \leq v \leq 2\pi, \quad \text{“Uniform distribution.”}$$

Furthermore, $\mathsf{E}[A^2]/2 = \sigma^2$, in accordance with the results in Section 2.4. The spectral density consists of two delta functions at $\pm f$ with weights $\sigma^2/2$. Adding independent cosines of the form (5.1) gives the following form.

Gaussian process in “random phase and amplitude” form

A Gaussian stationary, mean zero, process with discrete spectrum $\sigma_k^2/2$ at $\pm f_k, k = 1, \ldots, n \leq \infty$, and σ_0^2 at $f = 0$, has the representation

$$X(t) = X_0 + \sum_{k=1}^{n} \{X_k\cos(2\pi f_k t) + Y_k\sin(2\pi f_k t)\} \qquad (5.3)$$

$$= X_0 + \sum_{k=1}^{n} A_k\cos(2\pi f_k t + \phi_k), \qquad (5.4)$$

where X_0 and X_k, Y_k, are independent normal variables with $X_0 \sim N(0, \sigma_0^2)$, $\mathsf{E}[X_k] = \mathsf{E}[Y_k] = 0$, and $\mathsf{E}[X_k^2] = \mathsf{E}[Y_k^2] = \sigma_k^2$. The amplitudes $A_k = \sqrt{X_k^2 + Y_k^2}$ are Rayleigh distributed with $\mathsf{E}[A_k^2]/2 = \sigma_k^2$, and ϕ_k are uniformly distributed in $[0, 2\pi]$. All A_k and ϕ_k are independent.

The covariance function and spectral density are

$$r(\tau) = \sigma_0^2 + \sum_{1}^{n} \sigma_k^2\cos 2\pi f_\tau, \qquad (5.5)$$

$$R(f) = \sigma_0^2\,\delta_0(f) + \sum_{1}^{n} \frac{\sigma_k^2}{2}\big\{\delta_{-f_k}(f) + \delta_{f_k}(f)\big\}. \qquad (5.6)$$

5.3 The Wiener process

5.3.1 The one-dimensional Wiener process

The Wiener process (often instead called a Brownian motion) arguably is the most important of all stochastic processes. In Section 5.3.3 we outline parts of its history for the interested reader.

Definition 5.2 (Wiener process). *A Gaussian process* $\{X(t), 0 \leq t < \infty\}$ *is a Wiener process if* $X(0) = 0$, *and furthermore,*

a) *the increments* $X(t_2) - X(t_1), X(t_3) - X(t_2), \ldots, X(t_n) - X(t_{n-1})$, *are independent for all* $0 < t_1 < t_2 < \cdots < t_n$,

b) $X(t+h) - X(t)$ *has a normal distribution,* $N(0, \sigma^2 h)$, *for* $0 \leq t < t+h$. *In particular,* $X(t)$ *has a normal distribution* $N(0, \sigma^2 t)$, *i.e.,* $V[X(t)] = \sigma^2 t$.

A standardized Wiener process has $\sigma^2 = 1$.

The reader should compare this definition with the definition of a Poisson process, Definition 3.2. Both processes have independent stationary increments, but the increments have different distributions.

A Wiener process has independent stationary "increments," which may be positive as well as negative. That the increments are independent means exactly what is said in (a), while stationary increments mean that the distribution of the increment over an interval $(t_1, t_2]$ only depends on the interval length, $t_2 - t_1$.

Since the Wiener process has stationary independent increments, and since $X(0) = 0$, $E[X(1)] = 0$, and $V[X(1)] = \sigma^2$, we conclude the following from Theorem 3.3.

Theorem 5.1. *The Wiener process has mean value function* $m(t) = 0$, *variance* $V[X(t)] = \sigma^2 t$, *and covariance function*

$$r(s,t) = \sigma^2 \min(s,t).$$

This implies that the Wiener process is neither weakly, nor strictly, stationary. However, it is a Gaussian-Markov process.

Example 5.3 ("Ornstein-Uhlenbeck process and the Wiener process"). There is a simple relation between the Ornstein-Uhlenbeck process, i.e., the Gaussian process in Example 5.1, and the Wiener process. If $\{W(t), t \geq 0\}$ is a standard Wiener process with $V[W(t)] = t$, and

$$X(t) = e^{-\alpha t} W(e^{2\alpha t}), \quad \alpha > 0, \tag{5.7}$$

then $\{X(t), t \in \mathbb{R}\}$ is a stationary Gaussian process with covariance function

$$r(\tau) = \sigma^2 e^{-\alpha|\tau|}.$$

Obviously, $\{X(t)\}$ is a Gaussian process, since linear combinations of $X(t)$-variables are also linear combinations of $W(t)$-variables, and hence normal. The covariance function is obtained by direct calculation: for $\tau > 0$ (with analogous calculations for $\tau < 0$), one has

$$\begin{aligned}
r(\tau) &= C[X(t), X(t+\tau)] = C[e^{-\alpha t} W(e^{2\alpha t}), e^{-\alpha(t+\tau)} W(e^{2\alpha(t+\tau)})] \\
&= e^{-\alpha(2t+\tau)} C[W(e^{2\alpha t}), W(e^{2\alpha(t+\tau)})] = e^{-\alpha(2t+\tau)} r_W(e^{2\alpha t}, e^{2\alpha(t+\tau)}) \\
&= e^{-\alpha(2t+\tau)} \sigma^2 e^{2\alpha t} = \sigma^2 e^{-\alpha\tau}. \qquad \blacktriangle
\end{aligned}$$

5.3.2   Self similarity

The Wiener process is non-differentiable and is locally very irregular, particularly at a very close inspection, and it has the peculiar property of being, in a certain sense, equally irregular, regardless of what magnification we use.

Example 5.4 ("Self similarity of the Wiener process"). If $\{W(t), 0 \leq t < \infty\}$ is a standardized Wiener process with $V[W(t)] = t$, and $\lambda > 0$ is a constant, then the process $\{\widetilde{W}(t) = \lambda^{-1/2} W(\lambda t), 0 \leq t < \infty\}$ is also a standardized Wiener process. Therefore, the realizations of the Wiener process can be magnified by an arbitrary factor – they look statistically similar to the original, after scale change by the square root of the time magnification. For example, if we magnify the start of a Wiener process near $t = 0, W(0) = 0$, by a factor 100, i.e., $\lambda = 0.01$, and multiply by $\sqrt{100} = 10$, we get a process with exactly the same statistical properties as the one we started with. Repeating the procedure, we see that every tiny piece of a Wiener process has the same erratic behavior. The reverse is also true, the large scale fluctuations of a Wiener process are similar to those on a normal scale. $\blacktriangle$

This property of the Wiener process is a special case of *self similarity*, defined as follows.

Definition 5.3. *A stochastic process* $\{X(t), 0 \le t < \infty\}$ *is called* self similar *with* index H *if, for all n and* $\lambda > 0$*, and* $t_1, \ldots, t_n$*, the random vector* $(\lambda^{-H}X(\lambda t_1), \ldots, \lambda^{-H}X(\lambda t_n))$ *has the same distribution as* $(X(t_1), \ldots, X(t_n))$*, that is, a stochastic process is self similar if a scale change with a factor* λ^{-H} *combined with a time scale change by a factor* λ *does not change the distribution of the process.*

As shown in the example above, the Wiener process is self similar with index $H = 1/2$. Self similar processes are often used as models for Internet traffic, with the assumption that there is no natural scale for the amount of data transmitted over the net. Also economic time series, insurance processes, and many others are successfully modeled as self similar processes; see [19].

There is a generalization of Example 5.3, on the relation between self similar and stationary processes: A process $\{Y(t), 0 \le t < \infty\}$ is self similar with index H if and only if $X(t) = e^{-H\lambda t}Y(e^{\lambda t})$ is stationary for all $\lambda > 0$.

5.3.3   Brownian motion

This section contains a digression for the interested reader about the history of the Wiener process and its role as a physical model for Brownian motion.

Wiener's construction

The Wiener process is one of the first stochastic processes that got a rigorous mathematical treatment. Norbert Wiener[2] derived what we now would call a probability measure on the space of continuous functions, and showed that it leads to a process with independent, stationary, and normally distributed increments, and furthermore, that such a process is everywhere nondifferentiable. This was a remarkable theoretical achievement in the development of stochastic process theory, and the year was 1923, ten years before Kolmogorov's fundamental existence theorem; Appendix C.

Wiener defined his process so that the realizations were elements in $\mathbb{C}$, the space of continuous functions. One can construct the same process starting from the finite dimensional distributions, but that requires some extra mathematical machinery that guarantees that the process is uniquely defined by its values on the rational numbers.

---

[2]Norbert Wiener, American mathematician and probabilist, 1894–1964.

A stochastic process with independent increments, such as the Wiener process, can never be differentiable, unless it is constant. The Poisson process, which also has independent increments, is piecewise constant with occasional jumps, and hence non-differentiable at the jump time points. The interested reader can find more on this, for example, in [16, 33].

Brownian motion, Einstein, and the existence of molecules

Wiener also presented his process as a model for *Brownian motion*. The Scottish botanist Robert Brown (1773-1858) was a pioneer in the use of the microscope for natural studies. In 1827, he noticed how pollen grains suspended in water seemed to be in continuous erratic motion. He also observed the same behavior in pure dust particles, thereby ruling out the hypothesis that the movements were an expression of particle "life."[3]

The Brownian motion later turned out to be the one physically observable fact that settled "the great atom debate" of the late 19th century: is nature an infinitely divisible continuum, as claimed by Wilhelm Ostwald,[4] or does it consist of individually identifiable particles, called molecules, as Ludwig Boltzmann advocated? For history of the Brownian motion, see [34].

In the famous article *Zur Theorie der Brownschen Bewegung*, from 1906 [18], Albert Einstein showed how the movements of colloidal particles suspended in a liquid could be explained as the result of collisions with the molecules in the liquid. He assumed that the molecules move at random, independently of each other, and that each collision causes a very small displacement of the particle. Einstein was later in his life very skeptical towards any sort of randomness in nature, but at this time he used statistical arguments to derive the distribution of the particle displacement during a time interval of length t. By reference to the diffusion equation, he found the probability density of the displacement in the x-direction to be

$$f(x,t) = \frac{1}{\sqrt{4\pi Dt}} e^{-x^2/(4Dt)},$$

i.e., normal with variance $2Dt$, where D is the diffusion coefficient.

What made Einstein's work so conclusive was that he could give a physical argument, relating the displacement variance to other physical quantities:

$$D = 2RT/Nf,$$

---

[3] Actually, the phenomenon had been observed and reported already in 1784, by the Dutch physiologist Jan Ingenhousz, 1730–1799.

[4] Wilhelm Ostwald, German physicist, 1853–1932.

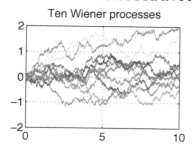
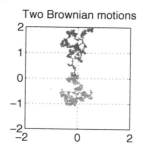

Figure 5.2 *Left: Ten realizations of a Wiener process with mean 0 and variance $t/12$. Right: Two Brownian motions with mean 0 and component variance $t/12$.*

where R is the ideal gas constant, T temperature in Kelvin, N the Avogadro number, i.e., the number of molecules in one mole of a substance, and f the friction coefficient. The Polish physicist Marian von Smoluchovsky (1872-1917), independently of Einstein and at the same time, obtained this result using quite different methods, [43].

The experiments that followed Einstein's and von Smoluchovsky's theoretical work finally verified the atom/molecular theory. The French physicist Jean-Baptiste Perrin (1870-1942) made very careful experiments with suspended rubber particles, and found complete agreement with the normal distribution, and, by estimating the variance $2Dt$ of the displacement, an approximate value of the Avogadro number N (*Die Atome*, 1914). The Swedish physicist The (Theodor) Svedberg (1884–1971) developed the ultracentrifuge and used it on small metal particles to verify the Einstein/Smoluchovsky theory (*Die Existens der Molekul* 1912).

Figure 5.2(a) shows ten realizations of a Wiener process with $E[X(t)] = 0$, $V[X(t)] = t/12$. Figure 5.2(b) shows two realizations of a Brownian motion in the plane, with independent vertical and horizontal displacements described by independent Wiener processes $X(t)$ and $Y(t)$ with $E[X(t)] = E[Y(t)] = 0$, $V[X(t)] = V[Y(t)] = t/12$. The diagram shows the displacement during 10 seconds of a particle starting at the origin.

5.4   Relatives of the Gaussian process

A linear transformation of a Gaussian process will give a new Gaussian process. Here are two simple non-linear transformations worth special study.

Example 5.5 ("Chi-squared process"). Let $\{X_k(t), t \in \mathbb{R}\}$, $k = 1, \ldots, n$, be independent, stationary Gaussian processes with mean 0 and variance 1. Then

$$Y(t) = \sum_{k=1}^{n} X_k^2(t)$$

defines a χ^2-process with n degrees of freedom. The marginal distribution of $Y(t)$ is a χ^2-distribution with probability density function

$$f_Y(y) = \frac{1}{2^{n/2}\Gamma(n/2)} y^{n/2-1} e^{-y/2}, \text{ for } y \geq 0.$$

In particular, for $n = 2$, the density is $f_Y(y) = \frac{1}{2}e^{-y/2}$, i.e., an exponential distribution with mean 2.   ▲

Example 5.6 ("Envelope"). In Section 8.3 we will define the *envelope* of a stationary process. It is a slowly varying process that traces the local variability of an oscillating process, like the band-limited process in Figure 5.1.

The envelope of a stationary Gaussian process $\{X(t)\}$ is a non-Gaussian process $\{Z_a(t)\}$, defined as

$$Z_a(t) = \sqrt{X^2(t) + Y^2(t)},$$

where $Y(t)$ is a stationary Gaussian process, the Hilbert transform of $X(t)$, defined in Section 8.3. The Y-process has the same covariance function as the X-process, and $X(t)$ and $Y(t)$ are uncorrelated, and hence independent, since $(X(t), Y(t))$ has a bivariate normal distribution.

However, for $s \neq t$, the variables $X(s)$ and $Y(t)$ are correlated, and hence dependent.

With $\sigma^2 = r_X(0) = V[X(t)] = V[Y(t)]$, the normalized squared envelope, $\sigma^{-2}Z_a^2(t) = \sigma^{-2}(X^2(t) + Y^2(t))$, is a χ^2-process with two degrees of freedom. This implies that the squared envelope is exponentially distributed with mean $2\sigma^2$, and probability density

$$f_{Z_a^2(t)}(x) = \frac{1}{2\sigma^2} e^{-x/(2\sigma^2)}, \ x \geq 0.$$

The density of the envelope $Z_a(t)$ is therefore

$$f_{Z_a(t)}(x) = 2x f_{Z_a^2(t)}(x^2) = \frac{x}{\sigma^2} e^{-x^2/(2\sigma^2)}, \text{ for } x \geq 0,$$

which is a Rayleigh distribution.   ▲

5.5 The Lévy process and shot noise process

5.5.1 The Lévy process

The Lévy processes are a class of processes related to the Wiener process and the Poisson process, studied in detail by Paul Lévy[5] and A.Y. Khinchin,[6] beginning around 1940. Initially a merely theoretical but very general notion, the Lévy process has now drawn considerable interest in mathematical finance. It combines periods of gradual continuous changes, with large and abrupt jumps of considerable size.

> **Definition 5.4.** *A stochastic process* $\{X(t), 0 \leq t < \infty\}$ *is said to be a Lévy process if it starts with* $X(0) = 0$ *and it furthermore has stationary and independent increments; cf. Definitions 3.2 and 5.2.*[7]

The Lévy process is a generalization of the Wiener process and also of the Poisson process. The Wiener process with an added drift term, $X(t) = W(t) + at$, is the only continuous Lévy process. The *compound Poisson process* is a Lévy process which is piecewise constant, and jumps at times determined by a Poisson process, but where the jumps are of random, independent size.

It turns out that any Lévy process is the sum of a Wiener process with drift and a countable number of independent compound Poisson processes with different jump intensities and jump distributions. Furthermore, for $\varepsilon > 0$, there are only a finite number of jumps with size greater than ε, and at most a countable number of jumps with size ε or smaller, in any finite time interval. If there are always only a finite number of jumps, it is a sum of a Wiener process with drift and a compound Poisson process.

5.5.2 Shot noise

Shot noise, in the physical meaning, is a type of irregular fluctuation in electronic or optical devices caused by the discrete character of electron or photon transports. The term "shot noise" is also used as an alternative name for *Poisson process with after-effects*, and it is used not only in physics, but also in econometrics, structural safety, and other areas.

---

[5]Paul Lévy, French mathematician and probabilist, 1886–1971.

[6]Aleksandr Yakovlevich Khinchin, Russian mathematician and probabilist, 1894–1959.

[7]In a more mathematical context, one also requires that the sample functions are continuous to the right and have limits from the left.

Example 5.7 ("Shot noise"). Electrons are emitted from the cathode in an electronic vacuum tube in an irregular stream, which can be described by a Poisson process with constant intensity λ. On its way towards the anode each electron generates an electric pulse of finite length. Different tubes may have different pulse shapes, but we assume different electrons in a tube to generate identical pulses; see figure 5.3 for some pulse shapes.

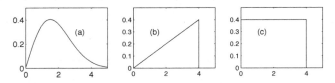

Figure 5.3 *Possible pulse shapes after electron emission.*

The total current at time t is the sum of currents caused by all electrons emitted before t. Since the emission times are random and there are only a finite number of pulses contributing to the current at any time, there will be small fluctuations – shot noise – in the current. The distribution of the process depends on the shape of the pulse, but if the intensity λ is very large, the distribution will be close to normal, and the shot noise process is approximately a Gaussian process. In optics, photons will cause a similar spatial effect, and with low light intensity, the distribution can be far from normal.   ▲

Mean and covariance function for shot noise, Campbell's formulas[8]

A shot noise process $\{X(t), t \in \mathbb{R}\}$ is the sum

$$X(t) = \sum_k g(t - \tau_k),$$

where $\{\tau_k\}$ are time points, generated by a Poisson process on the whole real axis, and $g(s)$ is an integrable function, the *impulse response.*

We shall derive the mean and covariance function of the shot noise process, but first we have to define the double sided Poisson process. It can be constructed from two independent Poisson processes with intensity λ, $\{N(t)\}$ and $\{N'(t)\}$, with jump times $\mu_1 < \mu_2 < \ldots$, and $\mu_0' < \mu_1' < \ldots$, respectively, where $\mu_1 > 0$ and $\mu_0' > 0$. The jump times $\{\tau_k\}_{k=-\infty}^{\infty}$, in the double sided Poisson process, can then be taken as

$$\tau_k = \begin{cases} \mu_k & \text{for } k \geq 1, \\ -\mu_{-k}' & \text{for } k \leq 0. \end{cases}$$

---

[8]Norman Robert Campbell, English physicist, 1880–1949.

We will present two methods to calculate the mean value function of $X(t)$. One way is to use that

$$E[X(t)] = E\left[\sum_k g(t - \tau_k)\right] = \sum_k E[g(t - \tau_k)], \qquad (5.8)$$

and calculate the exact expression for $E[g(t - \tau_k)]$. The other way is to use an approximation argument, which is very close to the original derivation by Campbell.

We start with the exact method. First, realize that, for $k \geq 1$, the event $\{\tau_k \leq u\}$ occurs exactly if $N(u) \geq k$. Since $N(u)$ has a Poisson distribution with expectation λu, we get the distribution function for τ_k, $k \geq 1$, and $u > 0$, by summing the Poisson probabilities, $P(N(u) = j) = e^{-\lambda u}(\lambda u)^j / j!$,

$$F_{\tau_k}(u) = P(\tau_k \leq u) = P(N(u) \geq k) = 1 - P(N(u) \leq k - 1)$$

$$= 1 - \sum_{j=0}^{k-1} e^{-\lambda u}(\lambda u)^j / j!,$$

and by differentiation we get the probability density function,

$$f_{\tau_k}(u) = \sum_{j=0}^{k-1} \lambda e^{-\lambda u}\frac{(\lambda u)^j}{j!} - \sum_{j=0}^{k-1} e^{-\lambda u}\frac{j\lambda^j u^{j-1}}{j!} = \lambda e^{-\lambda u}\frac{(\lambda u)^{k-1}}{(k-1)!}.$$

For $u < 0, k \geq 1$, the density is $f_{\tau_k}(u) = 0$. Hence, for $k \geq 1$,

$$E[g(t - \tau_k)] = \int g(t - u) f_{\tau_k}(u)\,du = \lambda \int_0^\infty g(t - u) e^{-\lambda u}\frac{(\lambda u)^{k-1}}{(k-1)!}\,du,$$

and analogous calculations for $k \leq 0$ gives

$$E[g(t - \tau_k)] = \lambda \int_{-\infty}^0 g(t - u) e^{-\lambda u}(\lambda u)^{-k} / (-k)!\,du.$$

Inserting this into (5.8), we get the expectation

$$E[X(t)] = \lambda \sum_{k=1}^\infty \int_0^\infty g(t - u) e^{-\lambda u}(\lambda u)^{k-1} / (k-1)!\,du$$

$$+ \lambda \sum_{k=-\infty}^0 \int_{-\infty}^0 g(t - u) e^{-\lambda u}(\lambda u)^{-k} / (-k)!\,du$$

$$= \lambda \sum_{k=0}^\infty \int_{-\infty}^\infty g(t - u) e^{-\lambda u}(\lambda u)^k / (k)!\,du$$

$$= \lambda \int_{-\infty}^\infty g(t - u) e^{-\lambda u} \sum_{k=0}^\infty (\lambda u)^k / (k)!\,du = \lambda \int_{-\infty}^\infty g(t - u)\,du,$$

where in the last step we used that $e^x = \sum_{k=0}^{\infty} x^k/k!$.

After these lengthy calculations we have arrived at the intuitively reasonable result that shot noise has mean value function $m(t) = \lambda \int_{-\infty}^{\infty} g(t-u)\,du = \lambda \int_{-\infty}^{\infty} g(u)\,du$, i.e., the mean value is equal to the average number of emissions per time unit times the total contribution from a single pulse.

The approximate method assumes that $g(s)$ is supported by a finite interval $[a,b]$, where it is uniformly continuous. Divide the real axis into small intervals, by a sequence of points, $t_k = k/N$, $k = \ldots, -2, -1, 0, 1, 2, \ldots$. Then

$$X(t) = \sum_k g(t-\tau_k) \approx \sum_n g(t-t_n) \cdot \#\{\tau_k \in (t_{n-1}, t_n]\}, \tag{5.9}$$

$$\mathsf{E}[X(t)] \approx \sum_n g(t-t_n) \cdot \mathsf{E}[\#\{\tau_k \in (t_{n-1}, t_n]\}] = \sum_n g(t-t_n) \cdot \lambda/N$$

$$\rightarrow \lambda \int g(u)\,du, \tag{5.10}$$

when $N \to \infty$. The approximation errors in (5.9) and (5.10) become negligible due to the uniform continuity. Similar calculations, based on bivariate distributions, will lead to the following formula for the covariance function,

$$r(s,t) = \mathsf{C}[X(s), X(t)] = \lambda \int_{-\infty}^{\infty} g(u)\,g(u+s-t)\,du.$$

Summary: (Campbell's formulas) The shot noise processe is a stationary process with mean $m(t) = \lambda \int_{-\infty}^{\infty} g(u)\,du$ and variance $\mathsf{V}[X(t)] = \lambda \int_{-\infty}^{\infty} g(u)^2\,du$, and its covariance function is

$$r(\tau) = \lambda \int_{-\infty}^{\infty} g(u)\,g(u-\tau)\,du. \tag{5.11}$$

Figure 5.4 shows the shot noise covariance function $r(\tau)$ when $g(s)$ is a rectangle with height 1 over the interval $(-a, a)$.

5.6   Monte Carlo simulation of a Gaussian process from spectrum

Simulation of a stationary process from its spectrum requires some extra consideration, compared to the rather straightforward simulation from the covariance structure described in Section 2.7. The complications come from the great flexibility in the spectrum that allows a mixture of low frequency components, sharp spectral details at arbitrary location, and a "white noise type," almost flat component up to high frequencies.

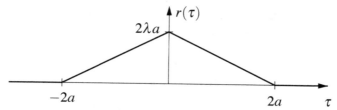

Figure 5.4 *Covariance $r(\tau)$ for shot noise with rectangular pulse.*

There are two different methods for simulation of a discrete sample of a Gaussian stationary process from a continuous spectrum. The first is to simply compute the Fourier transform of the spectrum, i.e., the covariance function, and then use covariance simulation by circulant embedding. The second method is to approximate the continuous spectrum by a discrete one and then generate a sample as a sum of cosine functions with random amplitudes and phases. The difference is not as big as it seems, since both methods normally make use of the Fast Fourier Transform (FFT) (see below).

Original simulation task: We are given a spectral density $R(f)$, $f \in \mathbb{R}$ for a stationary process $\{X(t), t \in \mathbb{R}\}$ in continuous time, and we want to generate a sample function $x(t)$, $0 \le t \le T$, of length T.

Discretization of time: For obvious reasons, we cannot hope to generate the continuous time function in a computer, so we have to decide on how dense a representation we want to generate. This means that we have to decide on a time step $\Delta_t > 0$ and generate $N = T/\Delta_t$ samples $x(t_k)$ with $t_k = k\Delta_t$ for $k = 0, 1, \ldots, N-1$; note that we let the indexing start with $k = 0$ so the last time point in the simulation will be $T - \Delta_t$.

Discretization of frequency: Simulation by means of the FFT-algorithm requires that also the spectrum is discrete, and that the number of samples and the number of frequencies are equal. We therefore approximate the spectrum by a discrete one over equally spaced frequencies $f_n = n\Delta_f$ for $n = 0, 1, \ldots, N-1$, for some small $\Delta_f > 0$. The highest frequency that contributes to the simulation will be $(N-1)\Delta_f$.

To discretize the spectrum we replace the (two-sided) spectral density $R(f)$ by a one-sided discrete spectrum with masses σ_n^2 at $f_n = n\Delta_f$,

$$\sigma_n^2 = \begin{cases} \Delta_f R(0), & n = 0, \\ 2\Delta_f R(f_n), & n = 1,\ldots,N-1. \end{cases} \tag{5.12}$$

The spectral mass σ_0^2 at $f = 0$ can normally be set to zero, since it represents a constant random average level for the entire simulation.

The choice of Δ_f depends on how complicated the spectral density is. It should be chosen small enough to make $\sum_n \sigma_n^2 \approx \int R(f)\,\mathrm{d}f = \mathsf{V}[X(t)]$. We further assume there is an upper frequency limit f_{max} above which the contribution to the process variance is negligible, and take N large enough to make $N\Delta_f > f_{max}$.

We have defined four parameters to be decided on for the simulation, T, N, Δ_t, Δ_f, and they are related in a very precise way. Before we state what these relations are, and explain why they are necessary, we describe the simulation algorithm.

5.6.1   Simulation by random cosines

The starting point for the simulation is the representation (5.3) of a Gaussian process with discrete spectrum. With $t = t_k = k\Delta_t$, and $f_n = n\Delta_f$, for $k = 0,\ldots,N-1$,

$$X(t_k) = \sum_{n=0}^{N-1} \sigma_n \{U_n \cos 2\pi f_n t_k + V_n \sin 2\pi f_n t_k\} \tag{5.13}$$

$$= \sum_{n=0}^{N-1} A_n \cos(2\pi f_n t_k + \phi_n), \tag{5.14}$$

with independent standard normal variables U_n, V_n, $A_n = \sigma_n\sqrt{U_n^2 + V_n^2}$, and $\cos\phi_n = \sigma_n U_n/A_n$, $\sin\phi_n = -\sigma_n V_n/A_n$.

One can simulate $X(t_k)$ from (5.13) by direct summation, without the FFT-algorithm. For small problems that is quite feasible. However, for large and repeated simulations this can be too time-consuming, and simulation can be very much speeded up if one uses the FFT. Then define

$$Z_n = A_n e^{-i\phi_n} = \sigma_n(U_n + iV_n), \tag{5.15}$$

and consider the complex sum

$$\sum_{n=0}^{N-1} Z_n e^{-i2\pi f_n t_k} = \sum_{n=0}^{N-1} A_n \cos(2\pi f_n t_k + \phi_n) - i\sum_{n=0}^{N-1} A_n \sin(2\pi f_n t_k + \phi_n). \tag{5.16}$$

The real part is precisely $X(t_k)$, and the imaginary part is the *Hilbert transform* of $\{X(t_k)\}$, with changed sign, to be described in Section 8.3.1, Example 8.7.

The Fast Fourier Transform

The fast Fourier transform (FFT) is a numerical scheme that enables very fast computation of the discrete Fourier transform z_k of a sequence Z_k,

$$z_k = \sum_{n=0}^{N-1} Z_n e^{-i2\pi nk/N}, \quad k = 0, \ldots, N-1. \tag{5.17}$$

For some details on how this is done, see Section 9.3.

To use the FFT for simulation with random Z_n as in (5.15), one has to make clear the relation between the number of sample points and the number of frequencies, (both are N), and the sampling interval Δ_t and the frequency interval Δ_f.

We identify, in (5.16) and (5.17), $f_n t_k = nk/N$ and get

$$\Delta_f \Delta_t = \frac{1}{N} = \frac{\Delta_t}{T},$$

since $T = N\Delta_t$. Hence, the fundamental restrictions

$$\Delta_f = \frac{1}{T}, \quad \Delta_t = \frac{T}{N}, \quad N\Delta_f \geq f_{max}. \tag{5.18}$$

Choice of simulation parameters

The simulation parameters depend on the intended simulation length and the desired resolution in time and/or frequency.

- $N\Delta_f \geq f_{max}$: This restriction is needed for the simulation to get correct variance. The highest frequency that contributes to the sample is $(N-1)\Delta_f$, so if $N\Delta_f$ is too small, the simulation will get too small variance. As an alternative, one can fold the spectrum around $\pm f_{max}$ and accept the aliasing that is introduced.

- $\Delta_f = 1/T$: This restriction is necessary to avoid that the simulated sequence repeats itself: $f_1 = \Delta_f$ is the smallest non-zero frequency in the sum (5.14), and the corresponding cosine function is periodic with period $1/\Delta_f$. Further, to resolve details in a complicated spectrum one needs a small Δ_f, which automatically leads to a large T, but also requires a large N to satisfy $N\Delta_f > f_{max}$.

- $\Delta_t = T/N$: Increasing N with constant T leads to better time resolution! One can therefore always increase the frequency range by *zero-padding*, i.e., adding a sequence of zeros in the upper end of the spectrum; see Section 9.3. Longer simulation time needs more terms in the sum, i.e., more frequencies. Zero-padding the spectrum to increase time resolution does not increase the variability in the result; it serves only to interpolate what is already there.

One should note the numerical efficiency of the FFT algorithm. The number of operations needed to calculate all z_k directly is of the order N^2, while the FFT with $N = 2^M$ needs of the order $N \log N$ operations; see also Section 9.3.

Simulation scheme

Thus, to simulate a sample $x_1, \ldots, x_N$ from a stationary, zero mean, Gaussian process with spectral density $R(f)$ one follows this scheme.

1. Decide on the maximum frequency f_{max},
2. Choose T or Δ_f, and the maximal time resolution Δ_t,
3. Use (5.18) and find Δ_f or T,
4. Calculate σ_n^2,
5. Pad with zeros to get $N = 2^M$ to satisfy all of (5.18),
6. Generate independent standard normal variables (U_n, V_n) and

$$Z_n = \sigma_n(U_n + iV_n),$$

7. Compute the sum (5.13) by direct summation, OR,
8. Compute the FFT, $z_k = FFT(Z_n)$ and take $x(t_k) = \Re(z_k)$.
9. (To generate the Hilbert transform, take $y(t_k) = -\Im(z_k)$.)

A ready-made procedure in MATLAB® for FFT-simulation from spectrum can be found in the MATLAB-toolbox WAFO, [48].

5.6.2 Simulation via covariance function

Simulation from a specified spectrum via the covariance function faces about the same considerations as direct simulation via cosine functions. If the spectrum is continuous one has to approximate it by lumping all energy to the discrete frequencies $f_k = k\Delta_f$ and then produce a covariance function

$$r(t_k) = \sum_n \sigma_k^2 \cos 2\pi f_n t_k,$$

according to (4.12). Generation of a random sample with this covariance function can then be done by means of circular embedding, as in Section 2.7. The same problems regarding maximum frequency, sampling interval, and frequency interval need attention, just as for the direct simulation method. We don't go into details, but refer to available routines, for example in the WAFO toolbox, [48].

Exercises

5:1. A Gaussian process $\{X(t), t \in \mathbb{R}\}$ has expectation zero and covariance function

$$r(\tau) = \begin{cases} 1 - |\tau|, & |\tau| \leq 1, \\ 0, & \text{otherwise.} \end{cases}$$

Compute $P(X(t) > 2)$, $P\left(X(t) + X\left(t + \frac{1}{2}\right) > 2\right)$, $P(X(t) + X(t+1) > 2)$.

5:2. Let $\{X(t), t \in \mathbb{R}\}$ be a Gaussian process with covariance function $r_X(\tau)$, and define the process $\{Y(t)\}$ by

$$Y(t) = X(t) - 0.4X(t-2).$$

Compute the covariance function of $\{Y(t)\}$. Is this process a Gaussian process? Is it stationary?

5:3. A Gaussian process $\{X(t), t \in \mathbb{R}\}$ has expectation zero, and covariance function

$$r_X(\tau) = \frac{2 + \tau^2}{1 + \tau^2}.$$

Determine the probability $P\left(X(2) > \frac{X(1) + X(3)}{2} + 1\right)$. Note that $r_X(\tau) \to 1$ as $\tau \to \infty$. How should you construct a process with this property, and what is its spectrum?

5:4. A disturbance in a communication system is modeled as a zero-mean Gaussian process $\{Z(t), t \in \mathbb{R}\}$ with the spectral density:

$$R_Z(f) = \frac{1}{1 + (2\pi f RC)^2}.$$

Determine the constant, RC, such that the instantaneous power of the disturbance is less than 1 with probability 95%. That is, determine RC, such that $P(Z^2(t) < 1) = 0.95$.

5:5. As you know, stock market prices fluctuate in a way that looks very much random. The following model has been used as a model of the price of a share stock. Let X_n be the price at day n. Define $Y_n = \log X_n$ with $Y_0 = 0$ and suppose that:

$$Y_{n+1} = Y_n + \log a + e_n,$$

where $a = 1 + p$, and p is the mean growth per day. Usually p is very small; for example, if the annual growth is 10%, then $p = 1.1^{1/250} - 1 \approx 0.00038$, if we count 250 business days per year. Suppose that e_n are independent normal variables with mean zero and variance $(K \log a)^2$.

(a) Compute the expectation, $E[Y_n]$, and the covariance function, $C[Y_m, Y_n]$. Is $\{Y_n\}$ a stationary process?

(b) It is reasonable to believe that $K = 30$. Compute the probability that the share price rises more than 3.8% (use $p = 0.00038$) from one day to another. What is the expected number of events of this type during one year (250 days)?

5:6. Let $\{Y(t), t \geq 0\}$ be a Wiener process with expectation $E[Y(t)] = 0$ and covariance function $r_Y(s,t) = \min(s,t)$. Define

$$Z(t) = \frac{Y(t) - Y(t/2)}{\sqrt{t}}, \quad t > 0.$$

(a) Determine the distribution of $Z(t)$.

(b) Is $\{Z(t), t > 0\}$ a weakly stationary process?

5:7. Let $\{X(t), t \in \mathbb{R}\}$ be a shot noise process with intensity $\lambda = 10^4$ and impulse response

$$g(t) = \begin{cases} 10^{-2} e^{-t}, & t > 0, \\ 0, & t \leq 0. \end{cases}$$

Determine approximately the probability $P(X(t+1) - X(t) > 2)$. Motivate the Gaussian process approximation.

Chapter 6

Linear filters – general theory

6.1 Introduction

A linear filter is an operation $\mathscr{S}$, that takes a function $x(t)$, the input, and transforms it into another function, $y(t) = (\mathscr{S}x)(t)$, the output, in such a way that the transform of a linear combination of functions is equal to the same linear combination of the transforms of the individual functions. Applied to stationary processes, they can be used for prediction of future values, interpolation of intermediate, missing values, and for identification of interesting features or removal of noise.

For prediction and interpolation, one wants to find a linear combination of known process values that approximates a not-observed future or intermediate value, as well as possible. To remove disturbing noise, for example from a music signal or from an image, one wants the filter to remove as much noise as possible, leaving the interesting part unaffected, or at least as close to the original as possible. If the task is to identify a feature, one would like to have the probabilities of making a wrong decision as small as possible.

Two new concepts, convolutions and generalized functions, will play a central role in this chapter. The *convolution* between two functions $\{h(t), t \in \mathbb{R}\}$ and $\{x(t), t \in \mathbb{R}\}$ is a new function

$$y(t) = \int_{-\infty}^{\infty} h(t-u)x(u)\,du = \int_{-\infty}^{\infty} h(u)x(t-u)\,du,$$

and the convolution between two sequences $\{h(t), t = 0, \pm 1, \ldots\}$ and $\{x(t), t = 0, \pm 1, \ldots\}$ is a new sequence

$$y(t) = \sum_{-\infty}^{\infty} h(t-u)x(u) = \sum_{-\infty}^{\infty} h(u)x(t-u).$$

The simplest example of a *generalized function* is a sum of an (ordinary) function and a delta function, but the concept of a generalized function is much more general than this; see Appendix B and [41]. (We sometimes write "ordinary function" when we want to emphasize that a function isn't a generalized function.)

Any linear filter which can act on all weakly stationary sequences can be expressed as a convolution. The situation for linear filters which act on processes in continuous time is more complicated. However, still a very wide class of linear filters in continuous time are those which can be expressed as a convolution with a generalized function. This class of filters contains all the filters of interest in this text, and is the only one we consider.

In Section 6.2 we treat the general methods for time-invariant filters, which are the most interesting for stationary process applications. We define the filter by its impulse response and frequency functions (frequency response), and use these to explain how the mean value, covariance, and spectral density functions are changed by the filter. Handling the covariance function is complicated, but the spectral relations are simple: the output spectral density is the input spectrum multiplied by the squared modulus of the filter frequency function.

Section 6.3 deals with infinite sums and other limiting operations, in particular differentiation and integration. Here we use a stochastic limit, the limit in quadratic mean (q.m.). The most obvious way to define the derivative of a stochastic process would be as the derivative of its sample functions. For mathematical reasons, we need a slightly less demanding derivative concept, namely as the q.m. limit of the difference quotient. Similarly, integrals are defined as q.m. limits of approximating Riemann sums. The moment functions of derivatives and integrals are obtained as the limits of the corresponding moments of the approximating difference quotients and sums.

In Section 6.4 we discuss white noise in continuous time. By this we mean a generalized stationary stochastic process, whose spectral density is constant over the whole real axis, and with the delta function as covariance function. We do not have the necessary mathematical machinery to give a complete definition, but we can present enough working rules for how to handle white noise as input to linear filters. The rules are formal, and have to be applied with some care.

In Section 6.5 we introduce a spectral tool to describe dependence between two stationary processes. The cross-spectral density is the Fourier transform of the cross-covariance function, defined in Section 2.2.1. It gives a complex-valued representation of the amplitude and phase relations between the processes, frequency by frequency.

Applications of the filter theory are given in the following two chapters. In Chapter 7 we handle the discrete time auto-regressive and moving average processes, while continuous time filters, including some simple stochastic differential equations, are treated in Chapter 8.

6.2   Linear systems and linear filters

A mathematical model for a physical dynamical system $\mathscr{S}$ describes how the state of the system develops under the influence of external factors, *input signals*, which change with time. The state of the system is described by a number of *output signals*, which describe the present state of the system. With one input signal $\{x(t)\}$ and one output signal $\{y(t)\}$, we may write $y = \mathscr{S}(x)$, meaning that the system transfers the input signal x to the output signal y.

A system is called *linear* if $\mathscr{S}(a_1 x_1 + a_2 x_2) = a_1 \mathscr{S}(x_1) + a_2 \mathscr{S}(x_2)$ for all constants a_1, a_2. Here we usually assume the constants a_1 and a_2 are real, but complex constants are sometimes also of interest. A system is called *time-invariant* if a time translation of the input signals causes the same time translation of the output signals. A system is called *stable*, if a bounded input signal is transformed into a bounded output signal. The mathematical formulation of a linear system is a *linear filter*.

Linear systems are the simplest and most useful of all mathematical models for complicated dynamical relations, partly because some physical phenomena indeed are linear, but also because of the mathematical simplicity of these models. Even if most physical phenomena are non-linear, the linear models are needed as necessary steps and building blocks in a realistic description. Examples of such building blocks in linear filters are differentiation, integration, addition, and translation..

6.2.1   Filter with random input

By a *stochastic dynamical system*, we mean a system where the input and output are stochastic processes. We will here assume that the system doesn't contribute any randomness, so that the same input always gives the same output. However, more general stochastic systems are also of interest. If the input to a linear and time-invariant filter is a stationary process, then also the output signal is stationary. We will now describe the relations between the statistical properties of the input and output signals, in particular covariance functions and spectral properties.

An example can explain why this is an important problem. Consider the suspension system for a car; cf. Example 1.6 in Chapter 1. In a simplified model, the system is completely described by the spring and damper constants, and in order to design a useful system with optimal properties one must know something about how the car is going to be used. Is it a car for easy cruising on motorways, a sports utility vehicle for combined terrain excursions and shopping of big items, or an environment friendly city hopper? Road holding and comfort will depend both on the suspension system and on the environment, road surface conditions, curvature, and on the driving conditions. Now, one cannot describe all possible roads and trips one is going to make with the car, but one can give some characteristic properties of these conditions in statistical terms, and some of the most important of these are the correlation and spectral properties of the road.

As examples of dynamical systems, let $\{X(t), t \in T\}$ be a stochastic process and define new processes $\{Y(t)\}$, by any of the following rules:

Translation: $Y(t) = X(t - c)$, for a constant c;

Differentiation: $Y(t) = X'(t)$;

Differential form: $Y(t) = a_0X(t) + a_1X'(t) + \cdots + a_pX^{(p)}(t)$;

Convolution form: Let $h(u)$ be a function, and define the output as a convolution integral or sum:

$$Y(t) = \int_{-\infty}^{\infty} h(t - u)X(u)\,du = \int_{-\infty}^{\infty} h(u)X(t - u)\,du,$$

$$Y(t) = \sum_{u=-\infty}^{\infty} h(t - u)X(u) = \sum_{u=-\infty}^{\infty} h(u)X(t - u). \tag{6.1}$$

These transformations are time-invariant; i.e., if they transfer the process $\{X(t)\}$ to the process $\{Y(t)\}$, then they transfer $\{X(t + c)\}$ to the process $\{Y(t + c)\}$, for every constant time shift c.

In order to deal properly with a relation like (6.1) we need to integrate a stochastic process, and compute expectations and covariances for integrals. Of course, this requires that the integrals exist, and they do, under appropriate conditions, for example if $h(u)$ is bounded and the process is continuous. We assume that all processes are regular enough for integrals to exist. More details can be found in Section 6.3, where we prove parts of the following rules, which hold as long as the integrals exist,

$$\mathsf{E}\left[\int_a^b g(t)X(t)\,dt\right] = \int_a^b g(t)\mathsf{E}[X(t)]\,dt = m_X \int_a^b g(t)\,dt, \tag{6.2}$$

$$\mathsf{C}\left[\int_a^b g(s)X(s)\,ds, \int_c^d h(t)X(t)\,dt\right] = \int_a^b \int_c^d g(s)h(t)r_X(t - s)\,dt\,ds. \tag{6.3}$$

6.2.2   Impulse response

Here we will consider a very wide class of linear time-invariant stochastic systems acting on weakly stationary processes, linear time invariant convolution filters. For simplicity we below omit "time invariant convolution" and just write "linear filter."

Definition 6.1. *a) In a continuous time linear filter the output stochastic process $\{Y(t)\}$ is obtained from the input process $\{X(t)\}$ through convolution with a generalized function $h(t)$,*

$$Y(t) = \int_{-\infty}^{\infty} h(t-u)X(u)\,du = \int_{-\infty}^{\infty} h(u)X(t-u)\,du.$$

b) In a discrete time linear filter the output stochastic process $\{Y(t)\}$ is obtained from the input process $\{X(t)\}$ through convolution with a sequence $h(t)$,

$$Y(t) = \sum_{-\infty}^{\infty} h(t-u)X(u)\,du = \sum_{-\infty}^{\infty} h(u)X(t-u)\,du.$$

The function $h(t)$ is called the impulse response of the system.

The impulse response is always assumed real in this book; complex forms are common in certain applications. That the impulse response is a generalized function makes it possible to express all the examples discussed above as linear filters. For example, the translation operation can be obtained by convolution with the delta function $\delta_c(u)$ located at c: this translates the input signal c time units,

$$Y(t) = \int \delta_c(u)X(t-u)\,du = X(t-c).$$

Similarly, to express differentiation in the form of an integral one just has to take the impulse response equal to the derivative of the delta function, $h(u) = \delta_0'(u)$. According to the "partial integration" rules for differentiation of the delta function (B.7) (note the sign),

$$Y(t) = \int_{-\infty}^{\infty} \delta'(u)X(t-u)\,du = -\int_{-\infty}^{\infty} \delta(u)\frac{d}{du}X(t-u)\,du = X'(t).$$

See Appendix B for details on the delta function and other generalized functions. For a full treatment, the interested reader is referred to [41, Ch. II].

If $h(t)$ only contains delta functions one can write the integral as a sum. All impulse functions in this section may contain delta functions, and we keep the notation as an integral throughout.

If the output signal $Y(t)$ only depends on input values $X(s)$ with $s \leq t$, then the filter is said to be *causal*. Mathematically, this is the same as requiring that $h(u) = 0$ for $u < 0$. For example, in discrete time,

$$Y(t) = \sum_{u=-\infty}^{\infty} h(u)X_{t-u} = \sum_{u=0}^{\infty} h(u)X_{t-u}.$$

A non-causal linear filter can often be approximated by a causal filter, if one can accept a short time delay between input and output.

Remark 6.1. *The name impulse response can be simply explained as the "response of an impulse": if the input signal $X(t)$ is an impulse at time 0, i.e., $X(t) = \delta_0(t)$, the output is equal to $h(t)$,*

$$Y(t) = \int_{-\infty}^{\infty} h(t-u)X(u)\,\mathrm{d}u = h(t).$$

6.2.3 Moment relations

From (6.2)–(6.3) one gets simple relations between the moment functions for a stationary input process and the corresponding functions for the output process in a linear time-invariant system.

Theorem 6.1. *If the input to a linear time-invariant system with impulse response $h(u)$ is a stationary process, then the output process (in continuous time),*

$$Y(t) = \int_{-\infty}^{\infty} h(u)X(t-u)\,\mathrm{d}u,$$

is also stationary. It has mean $m_Y = m_X \int_{-\infty}^{\infty} h(u)\,\mathrm{d}u$ and covariance function

$$r_Y(\tau) = \int_{-\infty}^{\infty}\int_{-\infty}^{\infty} h(u)h(v)r_X(\tau+u-v)\,\mathrm{d}u\,\mathrm{d}v. \tag{6.4}$$

In discrete time, $m_Y = m_X \sum_{u=-\infty}^{\infty} h(u)$, and

$$r_Y(\tau) = \sum_{u=-\infty}^{\infty}\sum_{v=-\infty}^{\infty} h(u)h(v)r_X(\tau+u-v). \tag{6.5}$$

In particular, the variance of the filter output is given by

$$V[Y(t)] = \begin{cases} \iint h(u)h(v)\, r_X(u-v)\, du\, dv, \\ \sum\sum h(u)h(v)\, r_X(u-v), \end{cases} \tag{6.6}$$

respectively.[1]

6.2.4 Frequency function and spectral relations

When a stationary process is transmitted through a time-invariant linear filter both amplitude and phases of the elementary components are affected in a way that is characteristic for the system, and defined by the *frequency function* of the filter.

Definition 6.2. *The function $H(f)$, defined as*

$$H(f) = \int_{-\infty}^{\infty} e^{-i2\pi f u} h(u)\, du, \quad \text{for } -\infty < f < \infty \text{ (continuous time)},$$

$$H(f) = \sum_{u=-\infty}^{\infty} e^{-i2\pi f u} h(u), \quad \text{for } -1/2 < f \le 1/2 \text{ (discrete time)},$$

is called the frequency function (or transfer function) of the filter with impulse response $h(u)$. It may be complex, $H(f) = |H(f)|e^{i\arg H(t)}$. Its modulus $|H(t)|$ is called the amplitude response (or the gain) and the argument $\arg H(f)$ is the phase response.

Example 6.1 ("Random phase and amplitude"). Consider the process

$$X(t) = A_0 + \sum_{k=1}^{n} A_k \cos(2\pi f_k t + \phi_k), \tag{6.7}$$

with pure cosine functions with independent random amplitudes $A_k > 0$, and phases ϕ_k, with $E[A_k^2] = 2\sigma_k^2$ as in (5.4). If $X(t)$ is the input signal to a linear filter with impulse response $h(u)$ and frequency function $H(f)$, so that

$$H(f) = |H(f)|e^{i\arg H(f)} = \int_{-\infty}^{\infty} e^{-i2\pi f u} h(u)\, du,$$

---

[1]Using convolution notation, i.e., $f * g(t) = \int f(u)g(t-u)\, du$, with $h_1(u) = h(-u)$, both (6.4) and (6.5) can be written $r_Y = h_1 * h * r_X$.

and

$$Y(t) = \int_{-\infty}^{\infty} h(t-u)X(u)\,du,$$

then, inserting (6.7), we get, after some calculation,

$$Y(t) = A_0 H(0) + \sum_{k=1}^{n} A_k |H(f_k)| \cdot \cos(2\pi f_k t + \phi_k + \arg H(f_k)).$$

This motivates the terms amplitude response for $|H(f)|$ and phase response for $\arg H(f)$. ▲

> **Important interpretation:** A linear filter with frequency function $H(f) = |H(f)|e^{i \arg H(f)}$ amplifies the amplitude of an elementary component with frequency f by a factor $|H(f)|$ and adds a phase shift $\arg H(f)$ to the phase, from ϕ to $\phi + \arg H(f)$.

The frequency function helps to simplify the relation between input and output covariance functions. Let $r_X(\tau)$ have the spectral representation

$$r_X(\tau) = \int e^{i2\pi f \tau} R_X(f)\,df,$$

where integration is over $(-\infty, \infty)$ or $(-1/2, 1/2]$, depending on if the process is in continuous or discrete time. Inserting this into (6.4), we get, with $\overline{H}$ denoting the complex conjugate of H,

$$r_Y(\tau) = \int\int h(u)h(v)\left\{\int e^{i2\pi f(\tau+u-v)} R_X(f)\,df\right\} du\,dv$$

$$= \int e^{i2\pi f\tau}\left\{\int h(u)e^{i2\pi fu}\,du \cdot \int h(v)e^{-i2\pi fv}\,dv\right\} R_X(f)\,df$$

$$= \int e^{i2\pi f\tau}\,\overline{H(f)}H(f)R_X(f)\,df = \int e^{i2\pi f\tau}|H(f)|^2 R_X(f)\,df. \quad (6.8)$$

Obviously, (6.8) is a spectral representation of the output covariance function with spectral density function $|H(f)|^2 R_X(f)$, as in the following theorem.[2]

---

[2]A simpler way to arrive at the spectral density is to use the convolution notation, $r_Y = h_1 * h * r_X$. Convolution of functions corresponds to multiplication of Fourier transforms, and hence $R_Y = \mathscr{F} r_Y = \mathscr{F} h_1 \times \mathscr{F} h \times \mathscr{F} r_X = \overline{H} \times H \times R_X = |H|^2 R_X$.

Theorem 6.2. *The relation between input and output spectral densities in a linear filter with impulse response function $h(u)$ and frequency function $H(f) = \int_{-\infty}^{\infty} h(u)e^{-i2\pi fu}\,du$ is*

$$R_Y(f) = |H(f)|^2 R_X(f).$$

The output mean and variance are, respectively,

$$m_Y = m_X H(0), \quad r_Y(0) = \int |H(f)|^2 R_X(f)\,df.$$

(The integration limits should be chosen according to if the process is in continuous or discrete time.)

In Theorem 6.1, we made a comment that stationarity is retained if a stationary stochastic input passes through a linear time-invariant filter. Furthermore, a Gaussian process is defined as a process where all linear combinations of process values have a normal distribution. Also limits of such linear combinations are normal. In a linear filter, the output is a limit of linear combinations of input variables, and if the input is Gaussian, then also the output is Gaussian. These invariance properties are illustrated in Figure 6.1.

Remark 6.2. *The notations $h_{XY}(u)$ and $H_{XY}(f)$ will sometimes be used to specify impulse response and frequency function for a filter with input $\{X(t)\}$ and output $\{Y(t)\}$.*

Example 6.2 ("Exponential smoothing, continuous time"). Let us consider a

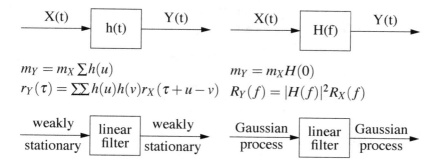

Figure 6.1 *Stationarity and normality are retained in a linear time-invariant filter.*

linear filter in continuous time with impulse response,

$$h(u) = \begin{cases} \beta e^{-\alpha u} & \text{for } u \geq 0, \\ 0 & \text{for } u < 0, \end{cases}$$

and accompanying frequency function,

$$H(f) = \int_{-\infty}^{\infty} e^{-i2\pi f u} h(u)\, du = \beta \int_{0}^{\infty} e^{-(\alpha + i2\pi f)u}\, du = \frac{\beta}{\alpha + i2\pi f}.$$

The output is an exponentially smoothed version of old values of the input,

$$Y(t) = \int_{-\infty}^{\infty} h(u)X(t-u)\, du = \beta \int_{0}^{\infty} e^{-\alpha u} X(t-u)\, du = \beta \int_{-\infty}^{t} e^{-\alpha(t-u)} X(u)\, du.$$

In the spectral density

$$R_Y(f) = |H(f)|^2 R_X(f) = \frac{\beta^2}{|\alpha + i2\pi f|^2} R_X(f) = \frac{\beta^2}{\alpha^2 + (2\pi f)^2} R_X(f),$$

one can see that high frequency components are moderated by the filter, since $|H(f)|^2$ has its maximum for $f = 0$ and damps down as f increases.   ▲

Example 6.3 ("RC-filter"). The RC-filter is an electronic device with one resistance R and one capacitance C, described in Figure 6.2. It acts as an exponential smoother. The input potential $X(t)$ between A_1 and A_2 and output potential $Y(t)$ between B_1 and B_2 are related through the equation $RCY'(t) + Y(t) = X(t)$, which can be solved with an initial value $Y(t_0)$ at a starting time t_0,

$$Y(t) = Y(t_0)\, e^{-(t-t_0)/(RC)} + \frac{1}{RC} \int_{t_0}^{t} e^{-(t-u)/(RC)} X(u)\, du. \qquad (6.9)$$

If the starting time t_0 lies in the far past, $t_0 \approx -\infty$, the solution to (6.9) is approximately a stationary process,

$$Y(t) \approx \frac{1}{RC} \int_{-\infty}^{t} e^{-(t-u)/(RC)} X(u)\, du, \qquad (6.10)$$

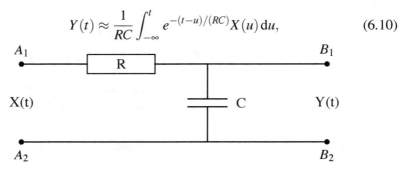

Figure 6.2 *Input potential $X(t)$ and output potential $Y(t)$ in an RC-filter.*

similar to the exponential smoother in Example 6.2 with $\alpha = \beta = 1/(RC)$.

If $\{X(t)\}$ is a weakly stationary input process, Equation (6.9) defines an output process $\{Y(t)\}$ with quite different properties. High frequencies are attenuated by a factor $|H(f)|^2 = \frac{1}{1+(RC)^2(2\pi f)^2}$, and the realizations of the Y-process will be much smoother than those of the X-process. For more on the RC-filter, see Example 8.2. ▲

Example 6.4 ("Exponential smoothing," discrete time). In the corresponding discrete time filter the impulse response is

$$h(u) = \begin{cases} (1-\theta)\theta^u & \text{for } u = 0,1,2,\ldots \\ 0 & \text{for } u = -1,-2,\ldots \end{cases}$$

where $\theta < 1$ is a smoothing constant, also called *forgetting factor*. The output is a geometrically smoothed version of the input,

$$Y(t) = (1-\theta)\sum_{u=0}^{\infty} \theta^u X(t-u) = (1-\theta)\sum_{u=-\infty}^{t} \theta^{t-u} X(u).$$

The frequency function of this filter is obtained as a geometric sum,

$$H(f) = (1-\theta)\sum_{u=0}^{\infty} e^{-i2\pi fu}\theta^u = \frac{1-\theta}{1-\theta e^{-i2\pi f}} = \frac{1-\theta}{1-\theta\cos 2\pi f + \theta i\sin 2\pi f}.$$

The spectral density of the output is

$$R_Y(f) = |H(f)|^2 R_X(f) = \frac{(1-\theta)^2}{(1-\theta\cos 2\pi f)^2 + \theta^2\sin^2 2\pi f} R_X(f)$$

$$= \frac{(1-\theta)^2}{1+\theta^2 - 2\theta\cos 2\pi f} R_X(f), \qquad (6.11)$$

for $-1/2 < f \le 1/2$. ▲

6.3 Continuity, differentiation, integration

In many of the examples we have presented so far, we have assumed that it is possible to integrate or differentiate a stochastic process with continuous time. From a practical point of view, there is usually no difficulty in doing so. A realization of a stochastic process is a function of time. In all examples the realizations have either been continuous (and sometimes also differentiable), or piecewise continuous with jump discontinuities, like the Poisson process or the random telegraph signal. However, as hinted in Remark 1.1 and in

Appendix C, there are mathematical difficulties to overcome if one wants to guarantee continuity of the realizations; see [33, Ch. 2]. To deal with these difficulties would lead too far away from the scope of this book, and we will instead use a simple convergence and continuity concept that can be defined and checked by means of the mean value and covariance functions.

In applications of the theory one usually does not need to worry about continuity of the sample functions. The situation becomes a little bit more complicated when it comes to first and higher order derivatives, as we shall see in Section 6.3.2.

6.3.1 Quadratic mean convergence in stochastic processes

In Appendix A.3, we define quadratic mean (q.m.) convergence of a sequence of random variables. We summarize the basic properties from Definition A.2 and Theorems A.5 and A.6, before we introduce our new continuity concept.

A sequence of random variables $\{X_n\}$ converges in quadratic mean to the random variable X (in symbols $X_n \overset{q.m.}{\to} X$), if

$$E[(X_n - X)^2] \to 0 \quad \text{when} \quad n \to \infty.$$

The Cauchy and Loève criteria, formulated in Section A.3, state that X_n converges in quadratic mean to some random variable, if and only if,

Cauchy criterion: $E[(X_m - X_n)^2] \to 0$,

Loève criterion: $E[X_m \cdot X_n] \to c$, a constant,

when m and $n \to \infty$, independently of each other.

> **Definition 6.3.** *a) A stochastic process $\{X(t)\}$ is called continuous in quadratic mean at time t if*
>
> $$X(t+h) \to X(t) \text{ in quadratic mean, when } h \to 0,$$
>
> *i.e., if $E[(X(t+h) - X(t))^2] \to 0$, when $h \to 0$.*
>
> *b) A stationary stochastic process $\{X(t)\}$ is everywhere continuous in quadratic mean, if*
>
> $$E[(X(t+h) - X(t))^2] = 2(r_X(0) - r_X(h)) \to 0,$$
>
> *when $h \to 0$, i.e., if the covariance function is continuous for $t = 0$.*

If the realizations of a weakly stationary stochastic process are continuous functions (with probability one), then the process is also continuous in quadratic mean, but the opposite does not necessarily hold. For example, the random telegraph signal in Example 3.7 has a continuous covariance function, $r(\tau) = \frac{1}{4}e^{-\alpha|\tau|}$, and is therefore continuous in quadratic mean, but its realizations have jumps and are not continuous functions. The explanation is that the jumps occur at random time points with continuous distribution. If t_0 is a time point, chosen beforehand, the probability of a jump of size 1 in the interval $[t_0, t_0 + h]$ is of the order h, and the expected "mean square jump" is also of order h, thus tending to 0 with h.

Example 6.5 ("Random phase and amplitude"). Now we can define the process

$$X(t) = A_0 + \sum_{k=1}^{\infty} A_k \cos(2\pi f_k t + \phi_k)$$

as a quadratic mean limit of the sum $X_n(t) = A_0 + \sum_{k=1}^{n} A_k \cos(2\pi f_k t + \phi_k)$. We use the Cauchy criterion to show the convergence of the sum, under the condition,

$$\sum_{k=0}^{\infty} \sigma_k^2 = E[A_0^2] + \sum_{k=1}^{\infty} \frac{1}{2} E[A_k^2] < \infty.$$

Since all terms in $X(t)$ are independent, for $m > n$,

$$E\left[(X_m(t) - X_n(t))^2\right] = E\left[\left(\sum_{k=n+1}^{m} A_k \cos(2\pi f_k t + \phi_k)\right)^2\right]$$

$$= \sum_{k=n+1}^{m} \frac{1}{2} E[A_k^2] = \sum_{k=n+1}^{m} \sigma_k^2,$$

which tends to 0 when $m > n \to \infty$. The Cauchy criterion gives that the sum converges in quadratic mean to some random variable, which we define to be the infinite sum.   ▲

6.3.2   Differentiation

In this section we will define quadratic mean differentiability of a stochastic process and give conditions for its existence in terms of the *second spectral moment*, expressed in terms of angular spectrum[3] and frequency spectrum as

$$m_2 = \int_{-\infty}^{\infty} \omega^2 \widetilde{R}_X(\omega)\, d\omega = \int_{-\infty}^{\infty} (2\pi f)^2 R_X(f)\, df.$$

---

[3]The name "spectral moment" comes from the angular frequency representation; cf. (4.6).

We will now present some of the important properties of the derivative of a stationary process. Differentiation is a linear operation, and its frequency function is $H(f) = i2\pi f$, as we shall see in Example 6.10 and 6.11. But first the definition.

Definition 6.4. *A stochastic process $\{X(t)\}$ is said to be differentiable in quadratic mean with derivative $\{X'(t)\}$, if*

$$\frac{X(t+h) - X(t)}{h} \to X'(t) \quad \text{in quadratic mean,}$$

when $h \to 0$, for all $t \in T$, i.e., if

$$\mathsf{E}\left[\left(\frac{X(t+h) - X(t)}{h} - X'(t)\right)^2\right] \to 0 \quad \text{when } h \to 0.$$

We compute the moment functions of the derivative of a stationary process, and prove an important theorem about the spectral distribution.

Theorem 6.3. *Let $\{X(t), t \in T\}$ be a weakly stationary process with covariance function $r_X(\tau)$. Then,*

(a) $\{X(t)\}$ is differentiable in quadratic mean if and only if $r_X(\tau)$ is twice differentiable for every τ. The derivative $\{X'(t)\}$ is weakly stationary with mean zero, $m_{X'} = 0$, and with covariance function and spectral density

$$r_{X'}(\tau) = -r_X''(\tau), \tag{6.12}$$

$$R_{X'}(f) = (2\pi f)^2 R_X(f). \tag{6.13}$$

(b) $r_X(\tau)$ is twice differentiable, and hence $\{X(t)\}$ is differentiable in quadratic mean, if and only if

$$\int_{-\infty}^{\infty} (2\pi f)^2 R_X(f)\, df < \infty.$$

(c) The derivative $\{X'(t)\}$ of a Gaussian process $\{X(t)\}$ is also a Gaussian process.

Proof. (a) If $h^{-1}(X(t+h)-X(t)) \to X'(t)$ in quadratic mean, then, according to Theorem A.4 in the appendix,

$$m_{X'} = \mathsf{E}[\lim_{h\to 0} h^{-1}(X(t+h)-X(t))] = \lim_{h\to 0} \mathsf{E}[h^{-1}(X(t+h)-X(t))]$$
$$= \lim_{h\to 0} h^{-1}(m_X - m_X) = 0.$$

Anticipating the existence, we use the notation $r_{X'}(\tau)$ (although we have not yet proved that $X(t)$ is differentiable and $X'(t)$ is stationary) and do the following calculations,

$$r_{X'}(\tau) = \mathsf{C}\left[\lim_{k\to 0}\frac{X(t+k)-X(t)}{k}, \lim_{h\to 0}\frac{X(t+\tau+h)-X(t+\tau)}{h}\right]$$
$$= \lim_{h\to 0}\lim_{k\to 0}\mathsf{E}\left[\frac{X(t+k)-X(t)}{k}\cdot\frac{X(t+\tau+h)-X(t+\tau)}{h}\right]$$
$$= \lim_{h\to 0}h^{-1}\lim_{k\to 0}\left\{\frac{r_X(\tau+h-k)-r_X(\tau+h)}{k}-\frac{r_X(\tau-k)-r_X(\tau)}{k}\right\}$$
$$= \lim_{h\to 0}\frac{-r'_X(\tau+h)+r'_X(\tau)}{h} = -r''_X(\tau),$$

where we first let $k \to 0$ and then $h \to 0$. Next, suppose that h and k go to 0 independently of each other. We have that

$$\mathsf{E}\left[\frac{X(t+k)-X(t)}{k}\cdot\frac{X(t+h)-X(t)}{h}\right]$$
$$= \frac{1}{hk}\left(r_X(h-k)-r_X(-k)-r_X(h)+r_X(0)\right). \quad (6.14)$$

Introducing a function $f(k,h)$ and two derivatives,

$$f(k,h) = r_X(h-k)-r_X(-k),$$
$$f'_1(k,h) = \frac{\partial}{\partial k}f(k,h) = -r'_X(h-k)+r'_X(-k),$$
$$f''_{12}(k,h) = \frac{\partial^2}{\partial h\partial k}f(k,h) = -r''_X(h-k),$$

we can use the *mean value theorem* to see that there are $\theta_1, \theta_2 \in (0,1)$ such that (6.14) is equal to

$$\frac{f(k,h)-f(0,h)}{hk} = \frac{f'_1(\theta_1 k,h)}{h} = \frac{f'_1(\theta_1 k,0)+hf''_{12}(\theta_1 k,\theta_2 h)}{h}$$
$$= f''_{12}(\theta_1 k,\theta_2 h) = -r''_X(\theta_2 h-\theta_1 k). \quad (6.15)$$

Since we have assumed that $r_X''(\tau)$ is continuous, this expression tends to $-r_X''(0)$ when $h, k \to 0$. Theorem A.4 implies that one can change the order of expectation and limit operation, and hence the Loève criterion is satisfied, and the process is differentiable in quadratic mean.

Moreover, one can show that if $r_X''(\tau)$ exists, it is possible to differentiate the spectral representation $r_X(\tau) = \int e^{i2\pi f\tau} R_X(f)\,df$ twice under the integral sign, giving the spectral representation

$$-r_X''(\tau) = \int e^{i2\pi f\tau} (2\pi f)^2 R_X(f)\,df.$$

Hence $\{X'(t)\}$ has the spectral density (6.13).

(b) Similar calculations as in the proof of part (a), (cf. [33, Ch. 2]) show that

$$\lim_{h \to 0} 2(1 - r_X(h))/h^2 = \int_{-\infty}^{\infty} (2\pi f)^2 R_X(f)\,df,$$

and that $r_X(\tau)$ is twice differentiable if and only if the right hand side is finite.

(c) The derivative is the result of a linear operation, which preserves Gaussianity. ∎

Example 6.6. We check the differentiability condition on the process $X(t) = A\cos(2\pi f_0 t + \phi)$, with covariance function $r_X(\tau) = \sigma^2 \cos 2\pi f_0 \tau$. Theorem 6.3 says that the process derivative $X'(t) = -(2\pi f_0)A\sin(2\pi f_0 t + \phi)$ has covariance function

$$r_{X'}(\tau) = -r_X''(\tau) = (2\pi f_0)^2 \sigma^2 \cos 2\pi f_0 \tau.$$

Since $X'(t) = -(2\pi f_0)A\cos(2\pi f_0 t + \phi - \pi/2)$, one can arrive at the same conclusion by noting that the phase $\phi - \pi/2$ is uniformly distributed over an interval with length 2π. ▲

Example 6.7 ("Ornstein-Uhlenbeck process and the Wiener process"). The covariance function $r(\tau) = \sigma^2 e^{-\alpha|\tau|}$ of an Ornstein-Uhlenbeck process $\{X(t), t \in \mathbb{R}\}$ is not differentiable at $\tau = 0$ and for the spectral density one has $\int (2\pi f)^2 R(f)\,df = \infty$. Hence, the process is not differentiable in quadratic mean; in fact, the sample functions are nowhere differentiable.

In Example 5.3 we found a relation between the Wiener process $\{W(t), 0 \le t < \infty\}$ and the Ornstein-Uhlenbeck process, namely $X(t) = e^{-\alpha t}W(e^{2\alpha t})$, and we can draw the conclusion that the Wiener process is not differentiable. ▲

Theorem 6.4. *If* $\{X(t)\}$ *satisfies the conditions in Theorem 6.3, then* $\{X(t)\}$ *and* $\{X'(t)\}$ *have a cross-covariance function,*

$$r_{X,X'}(t, t+\tau) = C[X(t), X'(t+\tau)] = r'_X(\tau). \qquad (6.16)$$

Proof. This is a direct consequence of Theorem A.4. Since $h^{-1}(X(t+h) - X(t)) \to X'(t)$ in quadratic mean,

$$
\begin{aligned}
r_{X,X'}(t, t+\tau) &= \lim_{h \to 0} E[X(t) \cdot h^{-1}(X(t+\tau+h) - X(t+\tau))] \\
&= \lim_{h \to 0} h^{-1}(r_X(\tau+h) - r_X(\tau)) = r'_X(\tau).
\end{aligned}
$$

Further, $r_X(\tau)$ is a symmetric function, so $r'_X(0) = 0$. ∎

Remark 6.3. *It follows from* (6.16) *that process and derivative taken at the same time,* $X(t)$ *and* $X'(t)$, *are uncorrelated:*

$$r_{X,X'}(t,t) = C[X(t), X'(t)] = r'_X(0) = 0.$$

Values taken at different time points $s \neq t$ *are, however, usually correlated. For example, if* $\tau > 0$ *is close to 0, then* $r'_X(\tau) < 0$ *and*

$$C[X(t+\tau), X'(t)] = r'_X(-\tau) = -r'_X(\tau) > 0.$$

The derivative at a certain time point and the process shortly after are quite naturally positively correlated; an upward slope should lead to higher values in the near future, see Figure 6.3.

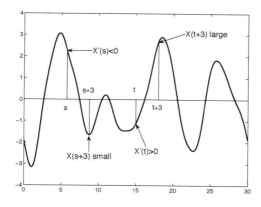

Figure 6.3 *Realization of a stochastic process. It illustrates the correlations between the derivative* $X'(t)$ *and the process* $X(t+\tau)$.

Theorem 6.5. *A stationary process $\{X(t), t \in T\}$ is n times differentiable in quadratic mean, if its covariance function $r_X(\tau)$ is 2n times differentiable. The cross-covariance function between derivatives is*

$$C[X^{(j)}(t), X^{(k)}(t+\tau)] = (-1)^{(j)} r_X^{(j+k)}(\tau),$$

for $j, k \leq n$. In particular, the process $\{X^{(n)}(t)\}$ has covariance function

$$r_{X^{(n)}}(\tau) = (-1)^n r_X^{(2n)}(\tau),$$

and spectral density

$$R_{X^{(n)}}(f) = (2\pi f)^{2n} R_X(f).$$

Proof. The proof follows the same line as that of Theorem 6.4, giving

$$
\begin{aligned}
r_{X^{(j)}, X^{(k+1)}}(\tau) &= C[X^{(j)}(t), X^{(k+1)}(t+\tau)] \\
&= \lim_{h \to 0} h^{-1}\left(r_{X^{(j)}, X^{(k)}}(\tau + h) - r_{X^{(j)}, X^{(k)}}(\tau)\right) = r'_{X^{(j)}, X^{(k)}}(\tau),
\end{aligned}
$$

$$
\begin{aligned}
r_{X^{(j+1)}, X^{(k)}}(\tau) &= C[X^{(j+1)}(t), X^{(k)}(t+\tau)] \\
&= \lim_{h \to 0} h^{-1}\left(r_{X^{(j)}, X^{(k)}}(\tau - h) - r_{X^{(j)}, X^{(k)}}(\tau)\right) = -r'_{X^{(j)}, X^{(k)}}(\tau).
\end{aligned}
$$

Differentiation of the first process in $C[X^{(j)}(t), X^{(k)}(t+\tau)]$ gives a minus sign in front of $r'_{X^{(j)}, X^{(k)}}(\tau)$, and differentiation of the second process gives a plus sign, which is the main content of the theorem.   ■

6.3.3   Integration

In Section 6.2.1 we presented rules for how to find mean and covariance between integrals. As seen, the formula for the covariance is quite cumbersome to use and should be avoided if possible. As promised, we can at least indicate the proof.

"Proof" of rule (6.2). We show that $E\left[\int_a^b g(t) X(t) \, dt\right] = m_X \int_a^b g(t) \, dt$. First, assume that the integration interval is bounded, $-\infty < a < b < \infty$, and let $a = t_0 < t_1 < \cdots < t_n = b$ be a subdivision of $[a, b]$ with $t'_i \in [t_{i-1}, t_i]$. One

has to prove that the Riemann-sum $S_n = \sum_{i=1}^{n} g(t_i')X(t_i')(t_i - t_{i-1})$ converges in quadratic mean, when the subdivision gets finer and finer. The limit, which will not depend on which sequence of subdivisions that has been chosen, is then denoted by $\int_a^b g(t)X(t)\,dt$.

The proof goes via the Loève criterion, applied to $E[S_m \cdot S_n]$, which under the stated condition, i.e., $\int\int g(s)g(t)E[X(s)X(t)]\,ds\,dt < \infty$, converges to a finite limit. According to Theorem A.4, the expectation is

$$E\left[\int_a^b g(t)X(t)\,dt\right] = E\left[\lim \sum g(t_i')X(t_i')(t_i - t_{i-1})\right]$$

$$= \lim \sum g(t_i')E[X(t_i')](t_i - t_{i-1}) = \lim \sum g(t_i')m_X(t_i - t_{i-1})$$

$$= m_X \int_a^b g(t)\,dt.$$

Formula (6.3) is proven in a similar way. ∎

6.4 White noise in continuous time

White noise is a useful generator of randomness and different kinds of stochastic processes. In discrete time, white noise is just a sequence of uncorrelated random variables with mean zero and constant variance. The spectral density is constant over $(-1/2, 1/2]$.

White noise in continuous time is not that simple. We would like to have similar properties as white noise in discrete time, namely,

(a) constant spectral density,

(b) process values are independent of each other.

Unfortunately, a stationary process with these properties would have infinite variance, and a physical realization would have infinite power. If the constant spectral density is $R_0 > 0$, the variance is

$$r(0) = V[X(t)] = \int_{-\infty}^{\infty} R_0\,df = \infty,$$

so it can not be a weakly stationary process (since these have all finite variance by definition).

One can, by Kolmogorov's existence theorem, construct a process of independent normal variables with finite variance, but that process would be useless for most purposes. For example, when it is passed through a linear filter with continuous impulse response function it would give zero output.

Furthermore, a realization of such a process would be everywhere discontinuous, and be unbounded in every finite interval. Its covariance function would be discontinuous at $\tau = 0$, and there would be no spectral representation.

To avoid the mentioned shortcomings, one can turn to stochastic processes which are *generalized functions* or *distributions*, in the mathematical sense. Such a process would have a "covariance function," which is also a distribution; for the white noise this would be the delta function. This covariance function corresponds to the property (b), namely $r(\tau) = R_0 \delta_0(\tau) = 0$, for $\tau \neq 0$, where $R_0 > 0$, is the constant "spectral density." The Fourier inversion formula

$$R_0 = \int e^{-i2\pi f\tau} R_0 \delta_0(\tau)\, d\tau$$

would hold, exactly as for ordinary covariance functions.

There is another nice property of this white noise: Gaussian white noise can be thought of as the derivative of (the non-differentiable) Wiener process, with derivation in the correct distributional meaning. The main application of this white noise is as input signal to a linear filter. If the impulse response is square integrable, $\int h(u)^2\, du < \infty$, the output signal will be an ordinary Gaussian stationary process!

Despite the mathematical difficulties involved in working with random processes where the realizations are generalized functions, there are many calculations where we can replace ordinary "almost white noise" with formal white noise with constant spectral density R_0 and covariance function $R_0 \delta_0(\tau)$.

Example 6.8 ("Almost white noise"). From Example 4.2 we know that the Ornstein-Uhlenbeck process $\{X(t), t \in \mathbb{R}\}$ with covariance function $r(\tau) = R_0 \frac{\alpha}{2} e^{-\alpha|\tau|}$ has the spectral density,

$$R(f) = \frac{R_0 \alpha}{2} \frac{2\alpha}{\alpha^2 + (2\pi f)^2} = R_0 \frac{1}{1 + (2\pi f/\alpha)^2}.$$

If α is very large, the spectral density is almost constant up to very high frequencies, and the covariance function $r(\tau)$ is almost zero, except for $\tau = 0$, where it is $R_0\alpha/2$. Furthermore, $\int_{-\infty}^{\infty} r(\tau)\, d\tau = R_0$ so the covariance function is almost like the scaled delta function $R_0 \delta_0$. Hence, for very large α, the process $\{X(t)\}$ is almost white noise in continuous time.

In practice, almost white noise can be physically realized in many different ways, all sharing the property that the covariance is almost zero for $\tau \neq 0$, and the spectral density is almost constant, $R_0 = \int_{-\infty}^{\infty} r(\tau)|d\tau$.   ▲

To be meaningful, calculations with white noise must be approximately true also for almost white noise. Used in linear filters, the result will usually be reasonable. For non-linear filters, one must be more careful, and use special rules, for example for differentiation. If one uses the formal (in the sense of generalized functions) derivative of the Wiener process, $W'(t)$, as Gaussian white noise, then the common derivation rule, $\frac{d}{dt} f(t)^2 = 2f(t)f'(t)$, does not hold. Instead $\frac{d}{dt} W(t)^2 = 2W(t)W'(t) + 1$, in distributional sense; see [60].

We summarize how to use white noise in continuous time.

- Make formal calculations with white noise as if it is a weakly stationary process with constant spectral density R_0 and covariance function $R_0 \delta_0(\tau)$.

- Check that the results are approximately true also for almost white noise. In linear filter problems, they usually are!

6.5 Cross-covariance and cross-spectrum

The covariance function $r(\tau)$ is a measure of the internal correlations within one stationary process, between $X(s)$ and $X(s + \tau)$ observed with time distance τ. The cross-covariance function, Definition 2.2 in Chapter 2, measures the correlation between two processes, $\{X(t)\}$ and $\{Y(t)\}$, as a function of the observation times, $r_{X,Y}(s,t) = C[X(s), Y(t)]$.

6.5.1 Definitions and general properties

Definition 6.5. *If the cross-covariance function $r_{X,Y}(s,t)$ only depends on the time difference $t - s$, the processes $\{X(t)\}$ and $\{Y(t)\}$ are said to be stationarily correlated, and the cross-covariance function is written*

$$r_{X,Y}(\tau) = C[X(t), Y(t + \tau)].$$

The matrix function

$$\mathbf{r}_{X,Y}(\tau) = \begin{pmatrix} r_X(\tau) & r_{X,Y}(\tau) \\ r_{Y,X}(\tau) & r_Y(\tau) \end{pmatrix}$$

is called the covariance matrix function.

Note that the order of X and Y is important in the definition of $r_{X,Y}(\tau)$, since it defines the covariance between the X-process and the Y-process observed τ time units later. Obviously,

$$r_{X,Y}(\tau) = r_{Y,X}(-\tau);$$

however, in most cases, the cross-covariance is not symmetric,

$$r_{X,Y}(\tau) \neq r_{X,Y}(-\tau).$$

Example 6.9. In Theorem 6.4, we learned that if $\{X(t)\}$ is stationary and differentiable, $\{X(t)\}$ and $\{X'(t)\}$ are stationarily correlated with cross-covariance function,

$$r_{X,X'}(\tau) = C[X(t), X'(t+\tau)] = r'_X(\tau),$$

so, indeed, $r_{X,X'}(-\tau) = -r_{X,X'}(\tau)$.   ▲

The cross-covariance function has a spectral representation, similar to that of the covariance function, $r_X(\tau) = \int e^{i2\pi f\tau} R_X(f)\,\mathrm{d}f$. We formulate the existence theorem, without proof.

Theorem 6.6. *Suppose $\{X(t)\}$ and $\{Y(t)\}$ are stationarily correlated processes with continuous covariance function matrix $\mathbf{r}_{X,Y}(\tau)$. Then there exists a spectral density matrix function*

$$\mathbf{R}_{X,Y}(f) = \begin{pmatrix} R_X(f) & R_{X,Y}(f) \\ R_{Y,X}(f) & R_Y(f) \end{pmatrix},$$

such that:
(a) In continuous time

$$r_{X,Y}(\tau) = \int_{-\infty}^{\infty} e^{i2\pi f\tau} R_{X,Y}(f)\,\mathrm{d}f.$$

The condition $\int |r_{X,Y}(\tau)|\,\mathrm{d}\tau < \infty$ is sufficient for the inversion formula $R_{X,Y}(f) = \int_{-\infty}^{\infty} e^{-i2\pi f\tau} r_{X,Y}(\tau)\,\mathrm{d}\tau$ to hold.

(b) In discrete time

$$r_{X,Y}(\tau) = \int_{-1/2}^{1/2} e^{i2\pi f\tau} R_{X,Y}(f)\,\mathrm{d}f.$$

The condition $\sum |r_{X,Y}(\tau)| < \infty$ is sufficient for the inversion formula $R_{X,Y}(f) = \sum_{\tau=-\infty}^{\infty} e^{-i2\pi f\tau} r_{X,Y}(\tau)$ to hold.

The cross-spectral densities $R_{X,Y}(f), R_{Y,X}(f)$ are complex-valued functions with

$$R_{X,Y}(-f) = \overline{R_{X,Y}(f)},$$
$$R_{Y,X}(f) = R_{X,Y}(-f).$$

The matrix $\mathbf{R}_{X,Y}(f)$ is of non-negative type, i.e., for every two complex numbers, $z_1, z_2,$

$$|z_1|^2 R_X(f) + z_1 \overline{z_2} R_{X,Y}(f) + \overline{z_1} z_2 R_{Y,X}(f) + |z_2|^2 R_Y(f) \geq 0,$$

and

$$0 \leq \frac{|R_{X,Y}(f)|^2}{R_X(f) R_Y(f)} \leq 1.$$

Example 6.10. The cross-spectrum between a stationary process $\{X(t)\}$ and its derivative $\{X'(t)\}$ is simple. If $\{X(t)\}$ has spectral density $R_X(f)$, then

$$r_{X,X'}(\tau) = r'_X(\tau) = \int e^{i2\pi f\tau}(i2\pi f) R_X(f)\, df,$$

so the cross-spectral density is $R_{X,X'}(f) = (i2\pi f) R_X(f)$. Figure 6.4 illustrates the cross-correlation $r_{X,X'}(\tau)$ for some choices of τ. Notice how $X(t)$ has negative correlation with $X'(t+2)$ and positive correlation with $X'(t+7)$. ▲

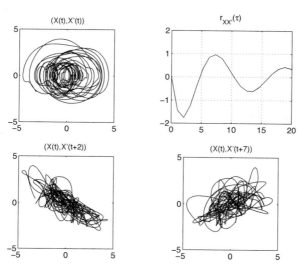

Figure 6.4 *Scatter plots of process values $X(t)$ (on the horizontal axis) and the derivative $X(t+\tau)$ (on the vertical axis) for $\tau = 0, 2, 7$. Upper right diagram shows the cross-covariance function $r_{X,X'}(\tau) = r'_X(\tau) = C[X(t), X'(t+\tau)]$. The process is Gaussian with a Jonswap spectrum; cf. Example 4.5.*

6.5.2   Input-output relations

Cross-covariance and cross-spectrum are simple tools to describe the statistical relations between a stochastic process and its derivative, as an example of an input-output relation in a linear filter. We will now study such relations in general, and also include the possibility that the output is disturbed by a stationary noise process $\{Z(t)\}$, so the general filter relation is

$$Y(t) = \int_{-\infty}^{\infty} h(t-u)X(u)\,du + Z(t) = \int_{-\infty}^{\infty} h(u)X(t-u)\,du + Z(t), \quad (6.17)$$

which is illustrated in this symbolic diagram:

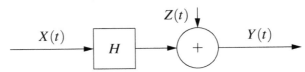

As is usual, we assume that $\{X(t)\}$ and $\{Z(t)\}$ are stationary processes and that the disturbance $\{Z(t)\}$ is uncorrelated with $\{X(t)\}$, i.e., $r_{X,Z}(s,t) = 0$, for all s and t. Then, $\{X(t)\}$ and $\{Y(t)\}$ are stationarily correlated and the following relations hold between the input and output signal characteristics.

Theorem 6.7. *Let $\{Y(t)\}$ be the output of the disturbed filter (6.17), with impulse response function $h(u)$ and frequency function $H(f) = \int_{-\infty}^{\infty} e^{-i2\pi fu} h(u)\,du$. Then the cross-covariance between input and output signal is*

$$r_{X,Y}(\tau) = \int_{-\infty}^{\infty} h(u) r_X(\tau - u)\,du, \qquad (6.18)$$

$$= \int_{-\infty}^{\infty} e^{i2\pi f\tau} H(f) R_X(f)\,df; \qquad (6.19)$$

and hence the cross-spectral density between input and output signal is

$$R_{X,Y}(f) = H(f) R_X(f). \qquad (6.20)$$

The spectral density of the output signal is

$$R_Y(f) = |H(f)|^2 R_X(f) + R_Z(f). \qquad (6.21)$$

Proof. Relation (6.18) follows directly, since $X(t)$ and $Z(t+\tau)$ are uncorrelated:

$$r_{X,Y}(\tau) = C\left[X(t), \int_{-\infty}^{\infty} h(u)X(t+\tau-u)\,du + Z(t+\tau)\right]$$

$$= \int_{-\infty}^{\infty} h(u)C[X(t),X(t+\tau-u)]\,du = \int_{-\infty}^{\infty} h(u)r_X(\tau-u)\,du.$$

The representation (6.19) follows from (6.18), if we use the spectral representation of $r_X(t)$:

$$r_{X,Y}(\tau) = \int_{-\infty}^{\infty} h(u)r_X(\tau-u)\,du = \int_{u=-\infty}^{\infty} h(u) \int_{f=-\infty}^{\infty} e^{i2\pi f(\tau-u)} R_X(f)\,df\,du$$

$$= \int_{f=-\infty}^{\infty} e^{i2\pi f\tau} R_X(f) \int_{u=-\infty}^{\infty} e^{-i2\pi fu} h(u)\,du\,df. \qquad \blacksquare$$

Remark 6.4. *The relation $R_{X,Y}(f) = H(f)R_X(f)$ is important, and can be used to estimate the frequency function $H(f)$ of a linear filter. An input stochastic process $X(t)$, with suitable frequency properties, is passed through the filter, giving an output $Y(t)$. By means of frequency analysis, described in Chapter 9, one then estimates the spectral density $R_X(f)$ of the input and the cross-spectrum $R_{X,Y}(f)$ between the input and output. The ratio $\widehat{H(f)} = R_{X,Y}(f)/R_X(f)$ is an estimate of the frequency function $H(f)$. It is important that the input contains enough power at the interesting frequencies.*

Remark 6.5. *Equation 6.17 and its schematic illustration is useful when the variations in the Y-process are directly caused by the X-process, except for the disturbances, of course.*

An appealing way to illustrate a common cause *for correlation between two processes $Y_1(t)$ and $Y_2(t)$ is the following:*

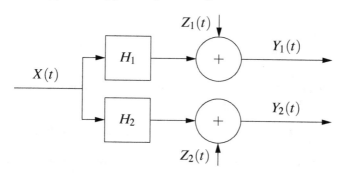

The covariance and spectral relations between $Y_1(t)$ and $Y_2(s)$ are found by the general formula for covariance between integrals, (6.3).

6.5.3   Interpretation of the cross-spectral density

Often it is more informative to consider and interpret the cross-spectral density $R_{X,Y}(f)$ than the cross-covariance function $r_{X,Y}(\tau)$, as an indicator of the relation between input and output in a filter. The covariance function is a summary of many different interaction factors, while the spectrum splits the relation between frequencies, in a way that can be given a physical meaning.

To facilitate interpretation we write the cross-spectrum in polar form,

$$R_{X,Y}(f) = A_{X,Y}(f)\, e^{i\Phi_{X,Y}(f)}, \qquad (6.22)$$

where $A_{X,Y}(f) \geq 0$ is the modulus of $R_{X,Y}(f)$, and $0 \leq \Phi_{X,Y}(f) < 2\pi$ the argument. By definition, $\Phi_{X,Y}(f)$ is always an ordinary function, while delta functions in $R_{X,Y}(f)$ are incorporated in $A_{X,Y}(f)$.

Definition 6.6. *In the representation* (6.22) *the modulus $A_{X,Y}(f)$ is called the cross-amplitude spectrum and the argument $\Phi_{X,Y}(f)$ is called the phase spectrum. The function*

$$\kappa_{X,Y}^2(f) = \frac{|R_{X,Y}(f)|^2}{R_X(f)R_Y(f)} = \frac{A_{X,Y}(f)^2}{R_X(f)R_Y(f)}$$

is called the squared coherence spectrum or the magnitude squared coherence.

Theorem 6.6 implies that the cross-amplitude spectrum is symmetric around $f = 0$, the phase spectrum is anti-symmetric, and the squared coherence is bounded by 1:

$$A_{X,Y}(-f) = A_{X,Y}(f),$$
$$\Phi_{X,Y}(-f) = -\Phi_{X,Y}(f),$$
$$0 \leq \kappa_{X,Y}^2 \leq 1.$$

Example 6.11. The cross-spectrum between a stationary process and its derivative is, according Example 6.10,

$$R_{X,X'}(f) = (i2\pi f)\, R_X(f) = 2\pi f\, R_X(f)\, e^{i\pi/2}.$$

This gives a frequency dependent cross-amplitude spectrum and constant phase spectrum,

$$A_{X,X'}(f) = 2\pi f\, R_X(f), \qquad \Phi_{X,X'}(f) = \pi/2.$$

Since $R_{X'}(f) = (2\pi f)^2 R_X(f)$, the squared coherence spectrum is constant,

$$\kappa^2_{X,X'}(f) = 1.$$

The interpretations of these relations are the following:

Phase spectrum constant and equal to $\pi/2$ means that each frequency component in $X(t)$ is also present in $X'(t)$, but with a phase shift of $\pi/2$.

Coherence spectrum equal to 1 means the amplitudes in $X'(t)$ are directly proportional to those in $X(t)$.

Cross-amplitude spectrum frequency dependent by a factor $2\pi f$ means that the proportionality factor is $2\pi f$.

Note that the squared coherence spectrum is always non-negative.   ▲

Remark 6.6. *If any of the spectra contain delta functions one has to be careful with the interpretation. If $A_{X,Y}(f)$ contains a delta function for some frequency f_0,*

$$A_{X,Y}(f) = A^{(1)}_{X,Y}(f_0)\delta_{f_0}(f) + A^{(2)}_{X,Y}(f),$$

then also

$$R_X(f) = R^{(1)}_X(f_0)\delta_{f_0}(f) + R^{(2)}_X(f),$$
$$R_Y(f) = R^{(1)}_Y(f_0)\delta_{f_0}(f) + R^{(2)}_Y(f),$$

and we can define

$$\kappa^2_{X,Y}(f_0) = \frac{A^{(1)}_{X,Y}(f_0)^2}{R^{(1)}_X(f_0)R^{(1)}_Y(f_0)}.$$

If $R_X(f)$ or $R_Y(f)$, but not $A_{X,Y}(f)$, contains a delta function for a frequency f_0, we define $\kappa^2_{X,Y}(f_0) = 0$.

Example 6.12. In Theorem 6.7 we considered the filter $Y(t) = \int_{-\infty}^{\infty} h(t - u)X(u)\,du + Z(t)$, with uncorrelated input $\{X(t)\}$ and noise $\{Z(t)\}$. Application of the theorem gives the following spectral relations:

$$R_Y(f) = |H(f)|^2 R_X(f) + R_Z(f),$$
$$R_{X,Y}(f) = H(f)R_X(f) = |H(f)|R_X(f)e^{i\arg H(f)},$$
$$A_{X,Y}(f) = |H(f)|R_X(f), \qquad (6.23)$$
$$\Phi_{X,Y}(f) = \arg H(f), \qquad (6.24)$$
$$\kappa^2_{X,Y}(f) = \frac{|H(f)|^2 R_X(f)}{|H(f)|^2 R_X(f) + R_Z(f)}. \qquad (6.25)$$

The equations (6.23)–(6.25) summarize the structure of the dependence.   ▲

Example 6.13 ("Random phase and amplitude"). Consider again the sum

$$X(t) = A_0 + \sum_{k=1}^{n} A_k \cos(2\pi f_k t + \phi_k) \qquad (6.26)$$

of cosine functions with independent amplitudes with $E[A_k^2] = 4b_k$, and uniform phases ϕ_k and use it as input to the filter defined by (6.17).

Now, $R_X(f) = \sum_{k=-n}^{n} b_k \delta_{f_k}(f)$, with $f_{-k} = -f_k$, and we get from Theorem 6.7 that

$$R_{X,Y}(f) = \sum_{k=-n}^{n} H(f_k) b_k \delta_{f_k}(f) = \sum_{k=-n}^{n} |H(f_k)| b_k \delta_{f_k}(f) e^{i \arg H(f_k)},$$

and hence

$$A_{X,Y}(f) = \sum_{k=-n}^{n} |H(f_k)| b_k \delta_{f_k}(f). \qquad (6.27)$$

Furthermore $R_Y(f) = |H(f)|^2 R_X(f) + R_Z(f)$, and if we assume that the noise spectrum $R_Z(f)$ can be written as

$$R_Z(f) = \sum_{k=-n}^{n} R_Z^{(1)}(f_k) \delta_{f_k}(f) + R_Z^{(2)}(f),$$

where $R_Z^{(2)}$ does not contain any delta functions at frequencies $f_{-n}, \ldots, f_n$, we get the coherence spectrum as

$$\kappa_{X,Y}^2(f) = \begin{cases} \dfrac{|H(f_k)|^2 b_k}{|H(f_k)|^2 b_k + R_Z^{(1)}(f_k)}, & f = f_k, \ k = -n, \ldots, n, \\ 0, & f \neq f_k, \ k = -n, \ldots, n \end{cases} \qquad (6.28)$$

(if both numerator and denominator are 0, we define the ratio to be 0).

We see from (6.27) and (6.28) that, if the coherence spectrum is close to 1 for some frequency f_k, then $R_Z^{(1)}(f_k)/|H(f_k)|^2 b_k$ is close to 0, i.e., the output signal $\{Y(t)\}$ is, at least at that frequency, determined by the amplified input signal, and not so much by the noise $Z(t)$. If, on the other hand, $\kappa_{X,Y}^2(f_k)$ is close to 0 at f_k, then the disturbance is the dominating factor. We can also see, from (6.27), that a small value for $A_{X,Y}(f_k)$ can be caused either from a weak signal $\{X(t)\}$ or a high damping factor $|H(f_k)|$ at that frequency. ▲

Remark 6.7. *The cross-amplitude spectrum between two processes $\{X(t)\}$ and $\{Y(t)\}$ describes the frequency specific coupling between the processes.*

Similarly, but in a more complicated way, the phase spectrum $\Phi_{X,Y}(f)$ describes how the phases in the two processes are linked. In Example 6.1 the phase spectrum $\Phi_{X,Y}(f_k) = \arg H(f_k)$ gives the extra frequency specific phase shift between the X- and the Y-process. Generally, the phase spectrum is a kind of average phase shift at frequency f for that part of $\{Y(t)\}$ that can be explained from $\{X(t)\}$. If $\kappa^2_{X,Y}(f)$ is almost 0 that part is only a minute part of $\{Y(t)\}$. Therefore the phase spectrum is only of interest for frequencies where $\kappa^2_{X,Y}(f)$ is substantial.

Example 6.14 ("Heart rate variability"). Spectral analysis of heart rate variability (HRV) is a non-invasive method to assess autonomic cardiac regulation. The so-called high frequency band (HF), at around 0.25 Hz, is related to respiratory sinus arrhythmia (RSA), and for a healthy person the correlation with respiration is usually high. Figure 6.5 shows the recorded respiratory variation and HRV-signal. Just by comparing the curves we see that there seems to be a strong correlation. The estimated correlation coefficient is $\rho = r_{XY}(0) \approx -0.5$, which tells us that there is some negative correlation but from the plots one could expect the correlation to be even more negative.

However, if we study the estimated cross-amplitude spectrum and phase spectrum in Figure 6.5(c,d), we see that the correlation shows up for a frequency slightly above 0.2 Hz, where the cross-amplitude spectrum is high. At this frequency the phase spectrum is close to 3, i.e., π, which is a sign of negative correlation (hint: $\cos x = -\cos(x + \pi)$). This agrees with the negative sign of the estimated ρ, suggesting that the time domain measure ρ is dominated by what is going on at frequencies close to 0.2 Hz. The coherence spectrum in Figure 6.6 shows what we now expect, the coherence is close to one for the frequency slightly above 0.2 Hz. (The negative sign does not show up in the squared coherence spectrum.)   ▲

Exercises

6:1. Let the stationary Gaussian process, $\{X(t), t \in \mathbb{R}\}$, with expected value $m_X = 3$ and covariance function $r_X(\tau) = 1/(1 + \tau^2)$, be the input to a linear filter with the impulse response $h(u) = \delta_0(u) - \delta_1(u)$, where $\delta_c(u)$ denotes the delta function located at c. Compute the probability

$$P(Y(1) > 3 + Y(0)),$$

where $\{Y(t), t \in \mathbb{R}\}$ is the output of the filter.

6:2. Let a time continuous Gaussian process $\{X(t)\}$ with expectation 1 be input to a linear filter with frequency function $H(f)$. Denote the output

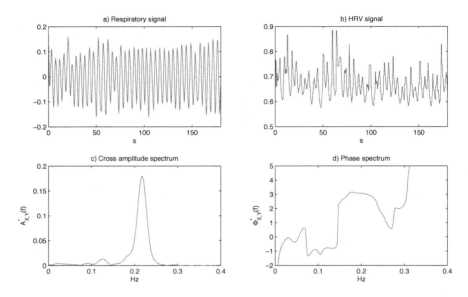

Figure 6.5 *Recordings of (a) respiratory variation and (b) Heart Rate Variability signal. (c) Estimated cross-amplitude spectrum and (d) phase spectrum.*

with $\{Y(t)\}$ and determine $\mathsf{P}[Y(t) \leq 4]$ if

$$R_X(f) = \begin{cases} 1 + |f| & \text{for } |f| \leq 1, \\ 0 & \text{for other values,} \end{cases} \qquad H(f) = \begin{cases} \frac{2}{1+|f|} & \text{for } |f| \leq 1, \\ 0 & \text{for other values.} \end{cases}$$

6:3. The input to a linear filter with frequency response $H(f) = i2\pi f/(1+$

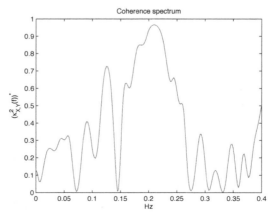

Figure 6.6 *Squared coherence spectrum of respiratory variation and HRV signal.*

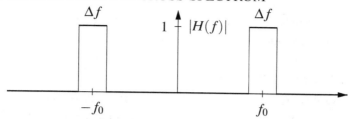

Figure 6.7 *An ideal bandpass filter.*

$i2\pi f$) is a stationary process $\{X(t)\}$ with mean zero and spectral density $R_X(f) = 1/(1+0.25(2\pi f)^2)$. Find the covariance function of the output.

6:4. A stationary process $\{X(t), t \in \mathbb{R}\}$ has mean zero and covariance function $r_X(\tau) = \max(0, 1 - |\tau|)$. Using this process as input signal to a linear filter, we can construct a new process, $\{Y(t), t \in \mathbb{R}\}$,

$$Y(t) = \int_{t-2}^{t} X(u)\,du.$$

(a) Compute the impulse response of this filter. Is the filter causal?
(b) What is the frequency response of the filter?
(c) Compute the spectral density of the output signal, $\{Y(t)\}$.

6:5. Suppose that a stationary process $\{X(t)\}$ with mean zero and spectral density $R_X(f)$ is processed by the ideal bandpass filter in Figure 6.7.

(a) Determine the average power of the output signal, $\{Y(t)\}$. Hint: The average power is $E\left[Y(t)^2\right]$.
(b) Suppose that we would like to estimate the energy content of a signal at frequency f_0. Explain how we can estimate $R_X(f_0)$, by using the filter, and by estimating the average power of the output signal. You may assume that $\{X(t)\}$ has a continuous spectrum and that $R_X(f)$ is a continuous function.

6:6. A time continuous random process, $\{X(t), t \in \mathbb{R}\}$, consists of a sum of three sinusoids:

$$X(t) = \sum_{k=1}^{3} A_k \cos(2\pi f_k t + \phi_k), \quad -\infty < t < \infty,$$

where $E[A_1^2] = 1$, $E[A_2^2] = 4$, $E[A_3^2] = 1$, and the phases ϕ_1, ϕ_2, and ϕ_3 are independent and uniformly distributed random variables in the interval $(-\pi, \pi]$. The frequencies are $f_1 = 4.5$, $f_2 = 7$, and $f_3 = 11$ kHz. The time continuous process is sampled. By mistake the sampling frequency is chosen $1/d = f_s = 5$ kHz.

(a) Which frequencies are present in the sampled process?

(b) The time discrete process is filtered with a filter with impulse response $h(n) = \frac{1}{2}\mathrm{sinc}(n/2)$. Which frequencies are present in the output signal after filtering? (Hint: $\mathrm{sinc}(x) = \frac{\sin(\pi x)}{\pi x}$.)

6:7. The weakly stationary process $\{X(t), t \in \mathbb{R}\}$ is differentiable and has the spectral density $R_X(f) = \pi e^{-2\pi|f|}$. Compute the covariance function and the spectral density of its derivative, $\{X'(t)\}$.

6:8. Prove that $r_X(\tau)$, defined by $r_X(\tau) = 4e^{-|\tau|/4} - 2e^{-|\tau|/2}$, can be a covariance function of a stationary process. Determine the spectral density and prove that the process is differentiable.

6:9. Two stochastic sequences $\{Y_n\}$ and $\{Z_n\}$ are generated by

$$Y_n = X_n + X_{n-1},$$
$$Z_n = 3X_n - 2X_{n-1} + X_{n-2},$$

where $\{X_n, n = 0, \pm 1, \pm 2, \ldots\}$ are independent random variables with mean 0 and variance σ^2. Compute the cross-covariance function between $\{Y_n\}$ and $\{Z_n\}$. At which time lag does the cross-covariance reach its maximum? What is the maximum?

6:10. Let $\{X(t)\}$ and $\{Z(t)\}$ be uncorrelated weakly stationary processes in continuous time and define $\{Y(t)\}$ by

$$Y(t) = \int_0^\infty h(s)X(t-s)\,ds + Z(t),$$

where $h(s)$ is the impulse response to a linear causal filter.

(a) Determine $r_{X,Y}(\tau)$ and $R_{X,Y}(f)$.

(b) What is the cross-covariance function and the cross-spectral density if the input signal, $\{X(t)\}$, is white noise?

Chapter 7

AR-, MA-, and ARMA-models

7.1 Introduction

This chapter deals with some of the oldest and most useful of all stationary process models, the *autoregressive* AR-model and the *moving average* MA-model. They form the basic elements in *time series analysis* of both stationary and non-stationary sequences, including model identification and parameter estimation. Predictions can be made in an algorithmically simple way in these models, and they can be used for efficient Monte Carlo simulation.

Modeling a stochastic process by means of a spectral density gives a lot of flexibility, since every non-negative, symmetric, integrable function is possible. On the other hand, the dependence in the process can take many possible shapes, which makes estimation of the spectrum or covariance function difficult. Simplifying the covariance, by assuming some sort of independence in the process generation, is one way to get a more manageable model. The AR- and MA-models both contain such independent generators. The AR-models include feedback, for example coming from an automatic control system, and they can generate sequences with strong dynamics. They are time-series versions of a *Markov model*. The MA-models are simpler, but less flexible, and are used when correlations have a finite time span.

George Udny Yule, British statistician (1871–1951), and Sir Gilbert Thomas Walker, British physicist, climatologist, and statistician (1868–1958), were the first to use AR-processes as models for natural phenomena. Yule in the 1920 suggested the AR(2)-process as an alternative to the Fourier method as a means to describe periodicities and explain correlation in the sunspot cycle, [56].

G.T. Walker was trained in physics and mathematics in Cambridge but worked for 20 years as head of the Indian Meteorological Department, where he was concerned with the serious problem of monsoon forecasting. He shared Yule's skepticism about deterministic Fourier methods, and favored correlation and regression methods. He made systematic studies of air pres-

165

sure variability at Darwin, Australia, and found that it exhibited a "quasi-periodic" behavior, with no single period, but rather a band of periods – we would say "continuous spectrum" – between 3 and $3\frac{1}{4}$ years. Walker extended Yule's AR(2)-model to an AR(p)-model and applied it to air pressure in Darwin, [50]. His name is now attached to the *Walker oscillation* as part of the complex El Niño – Southern Oscillation phenomenon.

In Section 7.2 we present the covariance and spectral theory for AR- and MA-processes, and also for a combined model, the ARMA-model. Section 7.3 deals with parameter estimation in the AR-model; for more on time series analysis, see [10, 11].

Section 7.4 contains a brief introduction to prediction methods based on AR- and ARMA-models. Under stationary conditions these methods can produce not only predictions but, for Gaussian models, also reliable bounds for the prediction errors. However, in many applications, models that allow more extreme and variable fluctuations are needed. One such model is the GARCH-model, which is the simplest model in a family of *non-linear models*, much used in financial applications. The model is presented in Section 7.5. A section on Monte Carlo simulation concludes the chapter.

7.2   Auto-regression and moving average

In this chapter, $\{e_t, t = 0, \pm 1, \dots\}$ denotes white noise in discrete time, i.e., a sequence of uncorrelated random variables with mean 0 and variance σ^2,

$$
\begin{aligned}
\mathsf{E}[e_t] &= 0, \\
\mathsf{C}[e_s, e_t] &= \begin{cases} \sigma^2 & \text{if } s = t, \\ 0 & \text{otherwise.} \end{cases}
\end{aligned}
$$

The sequence $\{e_t\}$ is called the *innovation process* with *innovation variance* σ^2. Its spectral density is constant,

$$
R_e(f) = \sigma^2 \text{ for } -1/2 < f \leq 1/2.
$$

7.2.1   *Auto-regressive process, AR(p)*

An auto-regressive process of order p, or shorter, AR(p)-process, is created by white noise passing through a feedback filter as in Figure 7.1.

An AR(p)-process is defined by its *generating polynomial* $A(z)$. Let $a_0 = 1, a_1, \dots, a_p$, be real coefficients and define the polynomial

$$
A(z) = a_0 + a_1 z + \dots + a_p z^p,
$$

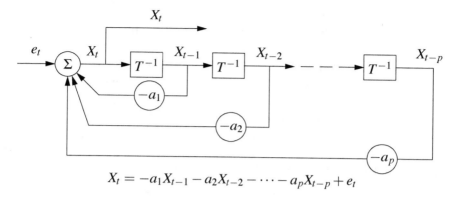

$$X_t = -a_1 X_{t-1} - a_2 X_{t-2} - \cdots - a_p X_{t-p} + e_t$$

Figure 7.1 *AR(p)-process. (The operator T^{-1} delays the signal one time unit.[1])*

in the complex variable z. The polynomial is called *stable* if the *characteristic equation*

$$z^p A(z^{-1}) = a_0 z^p + a_1 z^{p-1} \ldots + a_p = 0$$

has all its roots inside the unit circle, or equivalently, that all the zeros of the generating polynomial $A(z)$ lie outside the unit circle.

> **Definition 7.1.** *Let $A(z)$ be a stable polynomial of degree p. A stationary sequence $\{X_t\}$ is called an AR(p)-process with generating polynomial $A(z)$, if the sequence $\{e_t\}$, given by*
>
> $$X_t + a_1 X_{t-1} + \cdots + a_p X_{t-p} = e_t, \qquad (7.1)$$
>
> *is a white noise sequence, which is uncorrelated with $X_{t-1}, X_{t-2}, \ldots$. The variables e_t are the innovations to the AR-process. In a Gaussian stationary AR-process, also the innovations are Gaussian.*

Equation (7.1) becomes more informative if written in the form

$$X_t = -a_1 X_{t-1} - \cdots - a_p X_{t-p} + e_t,$$

where one can see how new values are generated as linear combinations of old value plus a small uncorrelated innovation. Note that it is important that

---

[1]$T^{-1}X_t = X_{t-1}, T^{-n}X_t = X_{t-n}.$

the innovation at time t is uncorrelated with the process so far; it should be a real "innovation," introducing something new to the process. Of course, e_t is correlated with X_t and all subsequent X_s for $s \geq t$.

If $A(z)$ is a stable polynomial of degree p, and $\{e_t\}$ a sequence of independent normal random variables, $e_t \sim N(0, \sigma^2)$, there always exists a Gaussian stationary AR(p)-process with $A(z)$ as its generating polynomial and e_t as innovations. The filter equation $X_t + a_1 X_{t-1} + \cdots + a_p X_{t-p} = e_t$ gives the X-process as solution to a linear difference equation. If the process was started a very long time ago the solution is approximately independent of the initial values; see Equation (7.18) in Theorem 7.5.

Theorem 7.1. *If $\{X_t\}$ is an AR(p)-process, with generating polynomial $A(z)$ and innovation variance σ^2, then $m_X = E[X_t] = 0$, and the spectral density is*

$$R_X(f) = \frac{\sigma^2}{\left|\sum_{k=0}^{p} a_k e^{-i2\pi f k}\right|^2} = \frac{\sigma^2}{|A(e^{-i2\pi f})|^2}.$$

The covariance function solves the Yule-Walker equations,

$$r_X(k) + a_1 r_X(k-1) + \cdots + a_p r_X(k-p) = 0, \text{ for } k = 1, 2, \ldots \quad (7.2)$$
$$r_X(0) + a_1 r_X(1) + \cdots + a_p r_X(p) = \sigma^2, \text{ for } k = 0. \quad (7.3)$$

Proof. The filter equation (7.1) defines the e_t-sequence as the output of a filter with $\{X_t\}$ as input, and with $h_k = a_k$, $k = 0, 1, \ldots, p$ as impulse response. The frequency function of that filter is $H(f) = A(e^{-i2\pi f}) = \sum_{k=0}^{p} a_k e^{-i2\pi f k}$, and according to Theorem 6.2, the relation between the spectral density $R_e(f) = \sigma^2$ for the innovations and $R_X(f)$ for the AR-process is

$$R_e(f) = |H(f)|^2 R_X(f),$$

$$R_X(f) = \frac{1}{|H(f)|^2} R_e(f) = \frac{\sigma^2}{\left|\sum_{k=0}^{p} a_k e^{-i2\pi f k}\right|^2}. \quad (7.4)$$

(Since all zeros of the generating polynomial $A(z)$ lie outside the unit circle, $H(f) \neq 0$.)

Further, taking expectations in (7.1), we find that $m_e = E[e_t]$ and $m_X = E[X_t]$ satisfy the equation

$$m_X + a_1 m_X + \cdots + a_p m_X = m_e,$$

i.e., $m_X A(1) = m_e = 0$, and since $A(1) \neq 0$, one has $m_X = 0$.

To show that $r_X(\tau)$ satisfies the Yule-Walker equations, we take covariances between X_{t-k} and the variables on both sides of Equation (7.1),

$$C[X_{t-k}, X_t + a_1 X_{t-1} + \cdots + a_p X_{t-p}] = C[X_{t-k}, e_t].$$

Here the left hand side is equal to

$$r_X(k) + a_1 r_X(k-1) + \cdots + a_p r_X(k-p),$$

while the right hand side is equal to 0 for $k = 1, 2, \ldots$, and equal to σ^2 for $k = 0$:

$$C[X_{t-k}, e_t] = \begin{cases} 0 & \text{for } k = 1, 2, \ldots \\ \sigma^2 & \text{for } k = 0. \end{cases}$$

This follows from the characteristics of an AR-process: For $k = 1, 2, \ldots$ the innovations e_t are uncorrelated with X_{t-k}, while for $k = 0$, we have

$$C[X_t, e_t] = C[-a_1 X_{t-1} - \cdots - a_p X_{t-p} + e_t, e_t] = C[e_t, e_t] = \sigma^2,$$

by definition.   ∎

The Yule-Walker equation (7.2) is a linear difference equation, which can be solved recursively. To find the initial values $r_X(0), r_X(1), \ldots, r_X(p)$ one has to solve the system of $p + 1$ linear equations,

$$
\begin{cases}
r_X(0) + a_1 r_X(-1) + \ldots + a_p r_X(-p) = \sigma^2, \\
r_X(1) + a_1 r_X(0) + \ldots + a_p r_X(-p+1) = 0, \\
\quad\quad\quad\quad\vdots \\
r_X(p) + a_1 r_X(p-1) + \ldots + a_p r_X(0) = 0.
\end{cases}
\tag{7.5}
$$

Note that there are $p + 1$ equations and $p + 1$ unknowns, $r_X(0), \ldots, r_X(p)$, since $r_X(-k) = r_X(k)$.

Remark 7.1. *There are (at least) three good reasons to use AR-processes in time series modeling:*

- *Many series are actually generated in a feedback system.*

- *The AR-process is flexible, and by smart choice of coefficients they can approximate most covariance and spectrum structures; and parameter estimation is simple.*

- *They are easy to use in forecasting: suppose we want to predict, at time t, the future value X_{t+1}, knowing all $\ldots, X_{t-p+1}, \ldots, X_t$. The linear predictor*

$$\widehat{X}_{t+1} = -a_1 X_t - a_2 X_{t-1} - \cdots - a_p X_{t-p+1}$$

is the best prediction of X_{t+1} in the least squares sense.

Example 7.1 ("AR(1)-process"). A process in discrete time with geometrically decaying covariance function is an AR(1)-process, and it can be generated by filtering white noise through a one-step feedback filter. With $\phi_1 = -a_1$, the recurrence equation

$$X_t + a_1 X_{t-1} = e_t, \quad \text{i.e.,} \quad X_t = \phi_1 X_{t-1} + e_t,$$

has a stationary process solution if $|a_1| < 1$. With $p = 1$, (7.5) becomes

$$r_X(0) + a_1 r_X(1) = \sigma^2,$$
$$r_X(1) + a_1 r_X(0) = 0,$$

which gives $V[X_t] = r_X(0) = \sigma^2/(1 - a_1^2)$. Then (7.2) gives the covariance function,

$$r(\tau) = \frac{\sigma^2}{1 - a_1^2} (-a_1)^{|\tau|} = \frac{\sigma^2}{1 - \phi_1^2} \phi_1^{|\tau|}.$$

The spectral density is

$$R(f) = \frac{\sigma^2}{|1 + a_1 e^{-i2\pi f}|^2} = \frac{\sigma^2}{1 + a_1^2 + 2a_1 \cos 2\pi f}$$

in accordance with Theorem 7.1.   ▲

Example 7.2 ("AR(2)-process"). An AR(2)-process

$$X_t + a_1 X_{t-1} + a_2 X_{t-2} = e_t$$

is a simple model for damped random oscillations with "quasi-periodicity," i.e., a more or less vague periodicity. The condition for stability is that the coefficients lie inside the triangle

$$\begin{aligned} |a_2| &< 1, \\ |a_1| &< 1 + a_2, \end{aligned}$$

illustrated in Figure 7.2.

The spectral density is, according to Theorem 7.1,

$$R(f) = \sigma^2 \left| 1 + a_1 e^{-i2\pi f} + a_2 e^{-i4\pi f} \right|^{-2}$$

$$= \frac{\sigma^2}{1 + a_1^2 + a_2^2 + 2a_1(1 + a_2)\cos 2\pi f + 2a_2 \cos 4\pi f}.$$

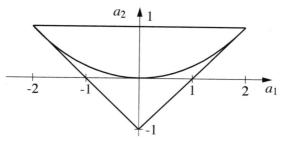

Figure 7.2 *Stability region for the AR(2)-process. The parabola $a_2 = a_1^2/4$ is the boundary between a covariance function of type (7.9) with complex roots, above the curve, and type (7.7) with real roots, below the curve.*

To find the variance and initial values in the Yule-Walker equation (7.2), we re-arrange (7.5), with $r(-k) = r(k)$, for $k = 1,2$:

$$\begin{cases} r(0) + \qquad a_1 r(1) + a_2 r(2) = \sigma^2, \\ a_1 r(0) + (1 + a_2) r(1) \qquad\qquad = 0, & (k = 1), \\ a_2 r(0) + \qquad a_1 r(1) + \quad r(2) = 0, & (k = 2), \end{cases}$$

leading to the variance and first two covariances,

$$r(0) = \frac{\sigma^2}{(1 + a_2)^2 - a_1^2} \cdot \frac{1 + a_2}{1 - a_2},$$

$$r(1) = -\frac{a_1}{1 + a_2} \cdot r(0), \qquad\qquad (7.6)$$

$$r(2) = \frac{a_1^2 - a_2 + a_2^2}{1 + a_2} \cdot r(0).$$

We can now express the general solution to the Yule-Walker equation in terms of the roots

$$z_{1,2} = -a_1/2 \pm \sqrt{(a_1/2)^2 - a_2},$$

to the characteristic equation, $z^2 A(z^{-1}) = z^2 + a_1 z + a_2 = 0$. The covariance function is of one of the types,

$$r(\tau) = K_1 z_1^{|\tau|} + K_2 z_2^{|\tau|}, \quad \text{when } a_1^2 > 4a_2, \qquad (7.7)$$

$$r(\tau) = K_1 z_1^{|\tau|}(1 + K_2|\tau|), \quad \text{when } a_1^2 = 4a_2, \qquad (7.8)$$

$$r(\tau) = K_1 \rho^{|\tau|} \cos(\beta|\tau| - \phi), \quad \text{when } a_1^2 < 4a_2, \qquad (7.9)$$

where the different types appear if the roots are (1) real-valued and different, (2) real-valued and equal, or (3) complex conjugated.

For the real root cases, the constants K_1, K_2, can be found by solving the equation system

$$\begin{cases} K_1 + K_2 & = r(0), \\ K_1 z_1 + K_2 z_2 = r(1), \end{cases}$$

with the starting values from (7.6).

For the complex root case, write the complex conjugated roots in polar form,

$$z_1 = \rho e^{i2\pi f} \quad \text{and} \quad z_2 = \rho e^{-i2\pi f},$$

where $0 < \rho < 1$ and $0 < f \le 1/2$. Then, the covariance function $r(\tau)$ is (for $\tau \ge 0$),

$$\begin{aligned} r(\tau) &= K_1 z_1^\tau + K_2 z_2^\tau = \rho^\tau (K_1 e^{i2\pi f\tau} + K_2 e^{-i2\pi f\tau}) \\ &= \rho^\tau \left((K_1 + K_2) \cos(2\pi f\tau) + i(K_1 - K_2) \sin(2\pi f\tau) \right) \\ &= \rho^\tau (K_3 \cos 2\pi f\tau + K_4 \sin 2\pi f\tau), \end{aligned}$$

where K_3 and K_4 are real constants (since $r(\tau)$ is real-valued). With

$$K_5 = |K_3 + iK_4| = \sqrt{K_3^2 + K_4^2}$$

and

$$\phi = \arg(K_3 + iK_4),$$

we can write

$$\begin{aligned} K_3 &= K_5 \cos \phi, \\ K_4 &= K_5 \sin \phi, \end{aligned}$$

and find that

$$r(\tau) = \rho^\tau K_5 \cos(2\pi f\tau - \phi).$$

Figure 7.3 shows realizations together with covariance function and spectral densities for two different Gaussian AR(2)-processes. Note, the peak in the spectral density for the process in (a), not present in (b), depending on the roots $z_{1,2} = -a_2/2 \pm \sqrt{(a_1/2)^2 - a_2}$.

We now discuss the two cases in Figure 7.3 in more detail.

Left case, complex roots: With $\sigma^2 = 1$, $a_1 = -1$, and $a_2 = 0.5$, the roots to the

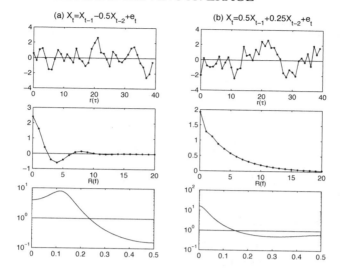

Figure 7.3 *Realization, covariance function, and spectral density (log scale) for two different AR(2)-processes with (a) $a_1 = -1$, $a_2 = 0.5$, and (b) $a_1 = -0.5$, $a_2 = -0.25$.*

characteristic equation are conjugate complex $z_{1,2} = (1 \pm i)/2 = 2^{-1/2}e^{\pm i\pi/4}$, and

$$X_t = X_{t-1} - 0.5X_{t-2} + e_t,$$

$$r_X(\tau) = \sqrt{6.4}\, 2^{-|\tau|/2} \cos(\frac{1}{4}\pi|\tau| - \theta), \text{ where } \theta = \arctan\frac{1}{3},$$

$$R(f) = (2.25 - 3\cos 2\pi f + \cos 4\pi f)^{-1}.$$

Note the weak peak in the spectrum near $f = 0.125 = (\pi/4)/(2\pi)$.

Right case, real roots: With $\sigma^2 = 1$, $a_1 = -0.5$, and $a_2 = -0.25$, the roots to the characteristic equation are real, $z_{1,2} = (1 \pm \sqrt{5})/4$, and we obtain, after some calculation,

$$X_t = 0.5X_{t-1} + 0.25X_{t-2} + e_t,$$

$$r(\tau) = \frac{0.96 + 0.32\sqrt{5}}{(-1 + \sqrt{5})^{|\tau|}} + \frac{0.96 - 0.32\sqrt{5}}{(-1 - \sqrt{5})^{|\tau|}},$$

$$R(f) = (1.32125 - 0.75\cos 2\pi f - 0.5\cos 4\pi f)^{-1}.$$

The possibility of having complex roots to the characteristic equation makes the AR(2)-process a very flexible modeling tool in the presence of "quasi-periodicities" near one single period. ▲

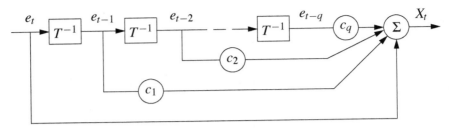

$$X_t = e_t + c_1 e_{t-1} + c_2 e_{t-2} + \cdots + c_q e_{t-q}$$

Figure 7.4 *MA(q)-process. (The operator T^{-1} delays the signal one time unit.)*

7.2.2   Moving average, MA(q)

A moving average process is generated by filtration of a white noise process $\{e_t\}$ through a transversal filter; see Figure 7.4.

An MA(q)-process is defined by its *generating polynomial*

$$C(z) = c_0 + c_1 z + \cdots + c_q z^q.$$

There are no restrictions on its zeros, like there are for the AR(p)-process, but it is often favorable that all zeros lie outside the unit circle; then the filter is called *invertible*. Expressed in terms of the *characteristic equation* $z^q C(z^{-1}) = 0$, the roots are then inside the unit circle. Usually, one normalizes the polynomial and takes $c_0 = 1$ and adjusts the innovation variance.

> **Definition 7.2.** *The process $\{X_t\}$, given by*
>
> $$X_t = e_t + c_1 e_{t-1} + \cdots + c_q e_{t-q},$$
>
> *is called a moving average process of order q, or, in abbreviated form, an MA(q)-process, with innovation sequence $\{e_t\}$ and generating polynomial $C(z)$.*

The sequence $\{X_t\}$ is an *improper* average as we do not require the weights c_k to be positive, and their sum need not be equal to 1.

Since the filter $X_t = \sum_{k=0}^{q} c_k e_{t-k}$ has impulse response function $h_k = c_k, k = 0, 1, \ldots, q$, and frequency function $H(f) = C(e^{-i2\pi f}) = \sum_{k=0}^{q} c_k e^{-i2\pi f k}$, the following theorem is a direct consequence of Theorems 6.1 and 6.2.

Theorem 7.2. *An MA(q)-process* $\{X_t\}$ *is stationary with* $E[X_t] = 0$,

$$r_X(\tau) = \begin{cases} \sigma^2 \sum_{j-k=\tau} c_j c_k, & \text{if } |\tau| \le q, \\ 0, & \text{otherwise}, \end{cases}$$

$$R_X(f) = \sigma^2 \left| \sum_{k=0}^{q} c_k e^{-i2\pi fk} \right|^2 = \sigma^2 |C(e^{-i2\pi f})|^2$$

$$= r_X(0) + 2 \sum_{\tau=1}^{q} r_X(\tau) \cos 2\pi f\tau.$$

Important: The main feature of an MA(q)-process is that its co-variance function is 0 for $|\tau| > q$.

Example 7.3. For the MA(1)-process, $X_t = e_t + c_1 e_{t-1}$ the covariance is

$$r(\tau) = \begin{cases} \sigma^2(1+c_1^2) & \text{for } \tau = 0, \\ \sigma^2 c_1 & \text{for } |\tau| = 1, \\ 0 & \text{for } |\tau| \ge 2, \end{cases}$$

and the spectral density is

$$R(f) = \sigma^2 \left| 1 + c_1 e^{-i2\pi f} \right|^2 = \sigma^2 \left((1 + c_1 \cos 2\pi f)^2 + c_1^2 \sin^2 2\pi f \right)$$
$$= \sigma^2 \left(1 + c_1^2 + 2c_1 \cos 2\pi f \right).$$

Figure 7.5 shows realizations, covariance functions, and spectral densities for two different Gaussian MA(1)-processes with $c_1 = \pm 0.9$, and

$$r(0) = 1.81, \quad r(1) = \pm 0.9,$$
$$R(f) = 1.81 \pm 1.8 \cos 2\pi f. \qquad \blacktriangle$$

7.2.3   Mixed model, ARMA(p,q)

A natural generalization of the AR- and MA-processes is a combination, with one AR- and one MA-filter in series, letting the right hand side of the

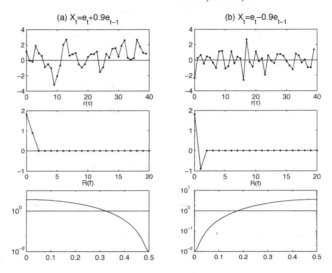

Figure 7.5 *Realizations, covariance functions, and spectral densities (log-scale) for two different MA(1)-processes: (a)* $X_t = e_t + 0.9e_{t-1}$, *(b)* $X_t = e_t - 0.9e_{t-1}$.

AR-definition (7.1) be an MA-process. The result is called an ARMA(p,q)-process,

$$X_t + a_1 X_{t-1} + \cdots + a_p X_{t-p} = e_t + c_1 e_{t-1} + \cdots + c_q e_{t-q}, \qquad (7.10)$$

where $\{e_t\}$ is a white noise process, such that e_t and X_{t-k} are uncorrelated for $k = 1, 2, \ldots$. It follows from Theorem 6.2 that $|A(e^{-i2\pi f})|^2 R_X(f) = |C(e^{-i2\pi f})|^2 R_e(f)$, and hence the spectral density is

$$R_X(f) = \sigma^2 \cdot \frac{|\sum_{k=0}^q c_k e^{-i2\pi fk}|^2}{|\sum_{k=0}^p a_k e^{-i2\pi fk}|^2} = \sigma^2 \cdot \frac{|C(e^{-i2\pi f})|^2}{|A(e^{-i2\pi f})|^2}. \qquad (7.11)$$

The covariance function of an ARMA(p,q)-process can be found via an extended Yule-Walker equation. We save the details until we have introduced some more mathematical tools; see Section 7.4.3.

Modeling ARMA-spectra by means of poles and zeros

The spectral density (7.11) of an ARMA-process is the ratio of the square of two complex polynomials in the argument $e^{-i2\pi f}$. This special form helps us to decide on suitable a- and c-coefficients in the ARMA-model. To see how it works, we look at the roots $z_1^a, \ldots, z_p^a$ of the characteristic equation $z^p A(z^{-1}) = 0$ of the AR-part, and $z_1^c, \ldots, z_q^c$ of the characteristic equation $z^q C(z^{-1}) = 0$ of the MA-part. These are called *poles* and *zeros*, respectively, of the system. Since the process is stationary, all poles lie inside the unit circle, and we assume that the same holds for the zeros, so the MA-part is invertible.

Now, take one of the polynomials, for example the MA-part and the numerator $|C(e^{-i2\pi f})|^2$ in (7.11). Since the C-polynomial has real coefficients, a root z_k^c is either real or is one in a complex conjugated pair, for example $(z_1^c, \overline{z}_1^c)$. Knowing the roots of the equation, we can factorize the polynomial,

$$z^q C(z^{-1}) = (z - z_1^c)(z - z_2^c) \cdot \ldots \cdot (z - z_q^c),$$

where the roots have been made "visible." The same operation for the AR-polynomial gives

$$z^p A(z^{-1}) = (z - z_1^a)(z - z_2^a) \cdot \ldots \cdot (z - z_p^a).$$

The spectral density is now obtained as a function of frequency f by inserting the argument $z = e^{i2\pi f}$, with $|z| = 1$, to give the following form of the spectral density $R_X(f)$:

$$\sigma^2 \frac{|z^q C(z^{-1})|^2}{|z^p A(z^{-1})|^2} = \sigma^2 \frac{|(e^{i2\pi f} - z_1^c)(e^{i2\pi f} - z_2^c) \ldots (e^{i2\pi f} - z_q^c)|^2}{|(e^{i2\pi f} - z_1^a)(e^{i2\pi f} - z_2^a) \ldots (e^{i2\pi f} - z_p^a)|^2}. \quad (7.12)$$

The factors in the numerator and denominator, e.g., $|e^{i2\pi f} - z_1^c|^2$ and $|e^{i2\pi f} - z_1^a|^2$, are the squared distances between the point $e^{i2\pi f}$ on the unit circle to the poles $z_1^a, z_2^a, \ldots, z_p^a$, and the zeros $z_1^c, z_2^c, \ldots, z_q^c$, all inside the unit circle. A complex pole z_k^a near the unit circle will generate a peak in the spectral density for $f \approx \pm \arg z_k^a$ and a zero at z_k^c will generate a dip for $f \approx \pm \arg z_k^c$. A real pole or zero will generate a peak or dip at $f = 0$ or $f = 1/2$, depending on the sign. For the AR-process it is the poles in the denominator of (7.12) that create the peaks in the spectrum. For an MA-process the zeros in the numerator produce dips. With a high order ARMA-model one can produce both strong peaks or even plateaus and bands of low spectral density.

Figure 7.6 shows an example of the product $|e^{i2\pi f} - z| \cdot |e^{i2\pi f} - \overline{z}|$ of the distances between $e^{i2\pi f}$ on the unit circle, and the complex conjugated points $z = (1+i)/2, \overline{z} = (1-i)/2$. The minimum distance is at $f \approx \pm 0.115$, slightly less than $\arg(1+i)/2 = 0.125$.

Figure 7.7 shows an example with an MA(4)-process with two pairs of complex zeros, one very close to the unit circle. As seen, the closer the zero is to the unit circle, the closer to zero is the spectral density at the specific frequency. Often the spectrum is shown in dB-scale, i.e., $10 \log_{10} R_x(f)$ is used to view the location of zeros.

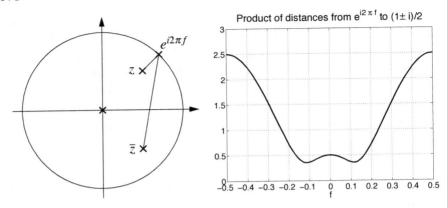

Figure 7.6 *Left: Distance from $e^{i2\pi f}$ to the complex conjugated $(z, \bar{z})$ determines the spectral density of an AR-process. Right: Product of the distances as function of frequency f.*

For the AR-process the distance products are in the denominator of the spectral density, and consequently, if the distance is small, the resulting spectral density (for this frequency) will be large. When the root is located at $\omega = 2\pi 0.125 = \pi/4$, the peak of the spectral density will be near $f = 0.125$; see Figures 7.6 and 7.8. When the roots are located closer to the unit circle,

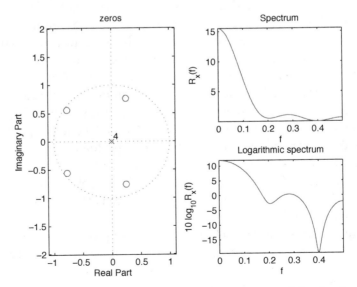

Figure 7.7 *Pole-zero plot, covariance function, and spectral density for an MA(4)-process. (The number 4 in the pole-zero plot indicates the order of the model.)*

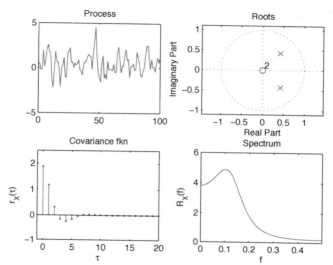

Figure 7.8 *One realization, the pole-zero plot, the covariance function, and the spectral density for an AR(2)-process.*

the peak in the spectral density is sharper; see Figure 7.9, where the poles now are located at $\omega = \pm 2\pi 0.25 = \pm \pi/2$ close to the unit circle and the corresponding spectral density has a peak near $f = 0.25$, which is much sharper than the one in Figure 7.8.

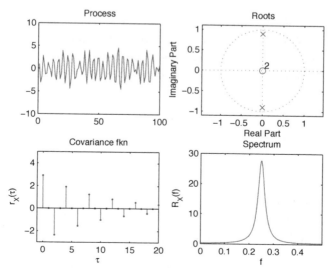

Figure 7.9 *Similar to Figure 7.8 but poles close to the unit circle.*

Remark 7.2 (Common roots). *If the A- and C-polynomials in an ARMA(p,q)-model have one or more zeros in common, i.e., some roots and poles coincide, then the corresponding factors in (7.12) cancel and $R_X(f)$ would be the spectrum of an ARMA-model of lower order. One could then reduce the number of parameters in (7.10). A common assumption in time-series analysis is therefore that no common factors exist.*

7.3 Estimation of AR-parameters

AR-, MA-, and ARMA-models are the basic elements in statistical time series analysis, both for stationary phenomena and for non-stationary. Here, we will only give a first example of parameter estimation for the most simple case, the AR(p)-process.

There are many methods for estimation of the parameters $a_1, \ldots, a_p$ and $\sigma^2 = V[e_t]$ in an AR(p)-process. The simplest, and often most reliable, way is to estimate the covariance function and then use the Yule-Walker equations to find suitable parameter values; see [11, Chapter 8].

Another way, which we will present here, is to use the auto-regressive structure

$$X_t + a_1 X_{t-1} + \ldots + a_p X_{t-p} = e_t, \tag{7.13}$$

which can be seen as a *multiple regression model*,

$$X_t = -a_1 X_{t-1} - \ldots - a_p X_{t-p} + e_t,$$

where the residuals (= innovations) e_t are uncorrelated with the regressors $X_{t-1}, X_{t-2}, \ldots$. With terminology borrowed from regression analysis one can call $U_t = (-X_{t-1}, \ldots, -X_{t-p})$ the *independent regressor variables*, and regard the new observation X_t as the *dependent* variable. With the parameter vector

$$\theta = (a_1, \ldots, a_p)',$$

we can write the AR-equation in standard multiple regression form,

$$X_t = U_t \theta + e_t,$$

and use standard regression techniques, as follows.

Suppose, we have n successive observations of the AR(p)-process (7.13), $x_1, x_2, \ldots, x_n$, and define $u_t = (-x_{t-1}, \ldots, -x_{t-p})$,

$$x_t = u_t \theta + e_t, \quad \text{for } t = p+1, \ldots, n.$$

The least squares estimate $\hat{\theta}$ is the θ-value that minimizes

$$Q(\theta) = \sum_{t=p+1}^{n} (x_t - u_t\theta)^2.$$

The solution can be formulated in matrix language. With

$$\mathbf{X} = \begin{pmatrix} x_{p+1} \\ x_{p+2} \\ \vdots \\ x_n \end{pmatrix}, \quad \mathbf{E} = \begin{pmatrix} e_{p+1} \\ e_{p+2} \\ \vdots \\ e_n \end{pmatrix},$$

$$\mathbf{U} = \begin{pmatrix} u_{p+1} \\ u_{p+2} \\ \vdots \\ u_n \end{pmatrix} = \begin{pmatrix} -x_p & -x_{(p-1)} & \cdots & -x_1 \\ -x_{(p+1)} & -x_p & \cdots & -x_2 \\ \vdots & \vdots & \vdots & \vdots \\ -x_{n-1} & -x_{n-2} & \cdots & -x_{n-p} \end{pmatrix},$$

the regression equation can be written in compact form, $\mathbf{X} = \mathbf{U}\theta + \mathbf{E}$, and the function to minimize is

$$Q(\theta) = (\mathbf{X} - \mathbf{U}\theta)'(\mathbf{X} - \mathbf{U}\theta).$$

Theorem 7.3. *The least squares estimates of the parameters* $(a_1, \ldots, a_p) = \theta$, *and the innovation variance* $\sigma^2 = \mathsf{V}[e_t]$, *in an AR(p)-process are given by*

$$\hat{\theta} = (\mathbf{U}'\mathbf{U})^{-1}\mathbf{U}'\mathbf{X}, \qquad \hat{\sigma^2} = Q(\hat{\theta})/(n-p).$$

The estimates are consistent, and converge to the true values when $n \to \infty$; *see Appendix A.4.*

The theorem claims that *if* the observations come from an AR(p)-process, then the parameters can be accurately estimated if the series is long enough. Also the covariance function and spectral density can then be estimated. However, one does not know for sure if the process is an AR(p)-process, and even if one did, the value of p would probably be unknown. Statistical time series analysis has developed techniques to test possible model orders, and to evaluate how well the fitted model agrees with data; for more on this, see [11].

Example 7.4. We use the technique on the AR(2)-process from Example 7.2, $X_t = X_{t-1} - 0.5X_{t-2} + e_t$. Based on $n = 512$ simulated values the estimates were amazingly close to the true values, namely $\widehat{a}_1 = -1.0361$, $\widehat{a}_2 = 0.4954$, $\widehat{\sigma} = 0.9690$; see Table 7.1 for the standard errors of the estimates.   ▲

	a_1	a_2	σ
true value	-1	0.5	1
estimated value	-1.0384	0.4954	0.969
standard error	0.0384	0.0385	

Table 7.1 *Parametric estimates of AR(2)-parameters.*

7.4   Prediction in AR- and ARMA-models

Forecasting or *predicting* future values in a time series is one of the most important applications of ARMA-models. Given a sequence of observations $x_t, x_{t-1}, x_{t-2}, \ldots$, of a stationary sequence $\{X_t\}$, one wants to predict the value $x_{t+\tau}$ of the process as well as possible in mean square sense, τ time units later. The value of τ is called the *prediction horizon*. We only consider predictors that are linear combinations of observed values.

7.4.1   *Prediction of AR-processes*

Let us first consider one-step ahead prediction, i.e., $\tau = 1$, and assume that the process $\{X_t\}$ is an AR(p)-process,

$$X_t + a_1 X_{t-1} + \cdots + a_p X_{t-p} = e_t, \tag{7.14}$$

with uncorrelated innovations $\{e_t\}$ with mean 0 and finite variance, $\sigma^2 = V[e_t]$, and with e_{t+1} uncorrelated with $X_t, X_{t-1}, \ldots$. In the relation (7.14), delayed one time unit,

$$X_{t+1} = -a_1 X_t - a_2 X_{t-1} - \cdots - a_p X_{t-p+1} + e_{t+1};$$

all terms on the right hand side are known at time t, except e_{t+1}, which in turn is uncorrelated with the observations of $X_t, X_{t-1}, \ldots$. It is then clear that it is not possible to predict the value of e_{t+1} from the known observations – we only know that it will be an observation from a distribution with mean 0 and variance σ^2. The best thing to do is to predict e_{t+1} with its expected value 0. The predictor of X_{t+1} would then be

$$\widehat{X}_{t+1} = -a_1 X_t - a_2 X_{t-1} - \cdots - a_p X_{t-p+1}. \tag{7.15}$$

Theorem 7.4. *The predictor (7.15) is optimal in the sense that if* $\widehat{Y}_{t+1}$ *is any other linear predictor, based only on* $X_t, X_{t-1}, \ldots$, *then*

$$E[(X_{t+1} - \widehat{Y}_{t+1})^2] \geq E[(X_{t+1} - \widehat{X}_{t+1})^2].$$

Proof. Since $\widehat{X}_{t+1}$ and $\widehat{Y}_{t+1}$ are based only on $X_t, X_{t-1}, \ldots$, they are uncorrelated with e_{t+1}, and since $X_{t+1} = \widehat{X}_{t+1} + e_{t+1}$, one has

$$
\begin{aligned}
E[(X_{t+1} - \widehat{Y}_{t+1})^2] &= E[(\widehat{X}_{t+1} + e_{t+1} - \widehat{Y}_{t+1})^2] \\
&= E[e_{t+1}^2] + 2E[e_{t+1}]E[\widehat{X}_{t+1} - \widehat{Y}_{t+1}] + E[(\widehat{X}_{t+1} - \widehat{Y}_{t+1})^2] \\
&= E[e_{t+1}^2] + E[(\widehat{X}_{t+1} - \widehat{Y}_{t+1})^2] \geq E[e_{t+1}^2] = E[(X_{t+1} - \widehat{X}_{t+1})^2],
\end{aligned}
$$

with equality only if $\widehat{Y}_{t+1} = \widehat{X}_{t+1}$. ∎

Repeating the one-step ahead prediction, one can extend the prediction horizon. To predict X_{t+2}, consider the identity

$$X_{t+2} = -a_1 X_{t+1} - a_2 X_t - \cdots - a_p X_{t-p+2} + e_{t+2},$$

and insert $X_{t+1} = \widehat{X}_{t+1} + e_{t+1}$, to get

$$
\begin{aligned}
X_{t+2} &= -a_1(-a_1 X_t - \cdots - a_p X_{t-p+1} + e_{t+1}) - a_2 X_t - \cdots - a_p X_{t-p+2} + e_{t+2} \\
&= (a_1^2 - a_2)X_t + (a_1 a_2 - a_3)X_{t-1} \cdots + (a_1 a_{p-1} - a_p)X_{t-p+2} \\
&\quad + a_1 a_p X_{t-p+1} - a_1 e_{t+1} + e_{t+2}.
\end{aligned}
$$

Here, $-a_1 e_{t+1} + e_{t+2}$ is uncorrelated with $X_t, X_{t-1}, \ldots$, and in the same way as before, we see that the best two-step ahead predictor is

$$
\begin{aligned}
\widehat{X}_{t+2} &= (a_1^2 - a_2)X_t + (a_1 a_2 - a_3)X_{t-1} \\
&\quad + \cdots + (a_1 a_{p-1} - a_p)X_{t-p+2} + a_1 a_p X_{t-p+1}.
\end{aligned}
$$

Repeating the procedure gives the best, in mean square sense, predictor for any prediction horizon.

7.4.2 *Prediction of ARMA-processes*

To predict an ARMA-process requires more work than the AR-process, since the unobserved old innovations are correlated with the observed data, and have delayed influence on future observations. An optimal predictor therefore requires reconstruction of old e_s-values, based on observed X_s, $s \le t$.

Let the ARMA-process be defined by

$$X_t + a_1 X_{t-1} + \cdots + a_p X_{t-p} = e_t + c_1 e_{t-1} + \cdots + c_q e_{t-q}, \qquad (7.16)$$

with $\{e_t\}$ a sequence of uncorrelated variables with mean 0 and variance σ^2. We present the solution to the one-step ahead prediction; generalization to many-steps ahead is very similar.

We formulate the solution by means of the generating polynomials (with $a_0 = c_0 = 1$),

$$A(z) = 1 + a_1 z + \cdots + a_p z^p,$$
$$C(z) = 1 + c_1 z + \cdots + c_q z^q,$$

where we assume that the polynomials have no common zeros.

With the backward translation operator T^{-1}, defined on page 167, $T^{-1} X_t = X_{t-1}$, $T^{-1} e_t = e_{t-1}$, etc., the defining equation (7.16) can be written in compact form as

$$A(T^{-1}) X_t = C(T^{-1}) e_t. \qquad (7.17)$$

The following representations are also useful.

> **Theorem 7.5.** *A stationary ARMA(p,q)-process $\{X_t, t \in \mathbb{Z}\}$ can be represented as an infinite moving average, an MA(∞)-process,*
>
> $$X_t = D(T^{-1}) e_t = \sum_{k=0}^{\infty} d_k e_{t-k}, \text{ with } d_0 = 1, \sum_0^{\infty} |d_k| < \infty. \qquad (7.18)$$
>
> *If the C-polynomial has all its zeros outside the unit circle, so that the MA-part is invertible, the ARMA(p,q)-process can be expressed as an infinite autoregression, an AR(∞)-process,*
>
> $$\sum_{k=0}^{\infty} f_k X_{t-k} = F(T^{-1}) X_t = e_t, \text{ with } f_0 = 1, \sum_0^{\infty} |f_k| < \infty. \qquad (7.19)$$

Proof. By assumption, the A-polynomial is stable and has all its zeros outside the unit circle. This implies that the quotient $C(z)/A(z)$ can be expanded in an infinite power series with convergence radius $z_0 > 1$,

$$\frac{C(z)}{A(z)} = 1 + d_1 z + d_2 z^2 + \ldots = \sum_0^\infty d_k z^k,$$

where $|d_k| < \text{constant} \cdot \theta^k$, for some $\theta < 1$. Replacing z by T^{-1} we obtain the formal series (7.18),

$$X_t = \frac{C(T^{-1})}{A(T^{-1})} e_t = \sum_0^\infty d_k e_{t-k}.$$

If the C-polynomial is invertible, i.e., has its zeros outside the unit circle, we can expand $A(z)/C(z)$ in a similar way to obtain (7.19).

The convergence of (7.18) and (7.19) when X_t and e_t are random variables needs some more arguments; see [11, Sec. 3.1]. ■

We now use the formal technique to find the best one-step ahead predictor for the ARMA(p,q)-process:

$$X_{t+1} = \frac{C(T^{-1})}{A(T^{-1})} e_{t+1} = e_{t+1} + \frac{C(T^{-1}) - A(T^{-1})}{A(T^{-1})T^{-1}} T^{-1} e_{t+1}$$

$$= e_{t+1} + \frac{C(T^{-1}) - A(T^{-1})}{A(T^{-1})T^{-1}} e_t. \tag{7.20}$$

According to (7.17) $e_t = \frac{A(T^{-1})}{C(T^{-1})} X_t$, and inserting this into (7.20) we get

$$X_{t+1} = e_{t+1} + \frac{C(T^{-1}) - A(T^{-1})}{C(T^{-1})T^{-1}} X_t.$$

Here, the innovation e_{t+1} is uncorrelated with known observations, while the second term only contains known X-values, and can be used as predictor. To find the explicit form, we expand the polynomial ratio in a power series,

$$\frac{C(z) - A(z)}{C(z)z} = \frac{(c_1 - a_1)z + (c_2 - a_2)z^2 + \cdots}{z(1 + c_1 z + \cdots + c_q z^q)} = g_0 + g_1 z + g_2 z^2 + \cdots,$$

which, with the T^{-1}-operator inserted, gives the desired form,

$$X_{t+1} = e_{t+1} + \frac{C(T^{-1}) - A(T^{-1})}{C(T^{-1})T^{-1}} X_t = e_{t+1} + g_0 X_t + g_1 X_{t-1} + g_2 X_{t-2} + \cdots.$$

Hence, the best predictor of X_{t+1} is

$$\widehat{X}_{t+1} = \frac{C(T^{-1}) - A(T^{-1})}{C(T^{-1})T^{-1}} X_t = g_0 X_t + g_1 X_{t-1} + g_2 X_{t-2} + \cdots. \tag{7.21}$$

Example 7.5. We show the simple calculations for an MA(1)-process $X_t = e_t + c_1 e_{t-1}$ with inverse $e_t = \frac{1}{1+c_1 T^{-1}} X_t$:

$$X_{t+1} = (1 + c_1 T^{-1}) e_{t+1} = e_{t+1} + c_1 e_t$$

$$= e_{t+1} + \frac{c_1}{1+c_1 T^{-1}} X_t = e_{t+1} + c_1 \sum_0^\infty (-c_1 T^{-1})^k X_t,$$

giving the one-step ahead predictor

$$\widehat{X}_{t+1} = c_1 \sum_0^\infty (-c_1)^k X_{t-k}. \qquad \blacktriangle$$

7.4.3   The covariance function for an ARMA-process

The MA(∞) series expansion in (7.18), $X_t = \sum_{k=0}^\infty d_k e_{t-k}$, can be used to give the initial values for a Yule-Walker equation for an ARMA-process. The coefficients d_j can be found by multiplying $A(z)D(z) = C(z)$ and equating coefficients for z^j on each side. With $a_0 = c_0 = 1$, we get successively

$$d_0 = 0,$$
$$d_1 = c_1 - d_0 a_1 = c_1 - a_1,$$
$$d_2 = c_2 - d_0 a_2 - d_1 a_1 = c_2 - a_2 - c_1 a_1 + a_1^2.$$

$$\cdots$$

Using this representation one can follow the track of proof for the AR-process and arrive at the following theorem; see further [11, Sec. 3.3].

Theorem 7.6 ("Yule-Walker equations for ARMA-models"). *The covariance function r_X for an ARMA(p,q)-process (7.10) is the solution to the Yule-Walker equation*

$$r_X(k) + a_1 r_X(k-1) + \cdots + a_p r_X(k-p) = 0,$$
$$\text{for } k \geq \max(p, q+1),$$

with the extended initial conditions

$$r_X(k) + a_1 r_X(k-1) + \cdots + a_p r_X(k-p) = \sigma^2 \sum_{j=k}^q c_j d_{j-k}, \quad (7.22)$$
$$\text{for } 0 \leq k < \max(p, q+1).$$

7.4.4 The orthogonality principle

The optimal AR-predictor (7.15) and Theorem 7.4 illustrate a general property of optimal linear prediction. Let Y and $X_1, \ldots, X_n$ be correlated random variables with mean zero.

> **Theorem 7.7.** *A linear predictor, $\widehat{Y} = \sum_{k=1}^n a_k X_k$, of Y by means of $X_1, \ldots, X_n$ is optimal in mean square sense if and only if the prediction error $Y - \widehat{Y}$ is uncorrelated with each X_k, so that $C[Y - \widehat{Y}, X_k] = 0$, for $k = 1, \ldots, n$. The coefficients $\{a_k\}$ of the optimal predictor are the solution to the linear equation system*
>
> $$C[Y, X_k] = \sum_{j=1}^n a_j C[X_j, X_k], \quad k = 1, \ldots, n. \qquad (7.23)$$

Example 7.6 ("Prediction of electrical power demand"). Trading electrical power on a daily or even hourly basis has become economically important and the use of statistical methods has grown. Predictions of demand a few days ahead is one of the basic factors in the pricing of electrical power that takes place on the power markets. Short term predictions are needed also for production planning of water generated power. (Of course, also forecasts of the demand several months or years ahead are important, but that requires different data and different methods.)

Electrical power consumption is not a stationary process. It varies systematically over the year, depending on season and weather, and there are big differences between days of the week, and time of the day. Before one can use any stationary process model, these systematic variations need to be taken into account. If one has successfully estimated and subtracted the systematic part, one can hope to fit an ARMA-model to the residuals, i.e., the variation around the weekly/daily profile. Figure 7.10 shows observed power consumption during one autumn week, together with one-hour ahead predicted consumption based on an ARMA-model, and prediction error. It turned out, in this experiment, that the ARMA-model contained just as much information about the future demand, as a good weather prediction, and that there was no need to include any more covariates in the prediction. ▲

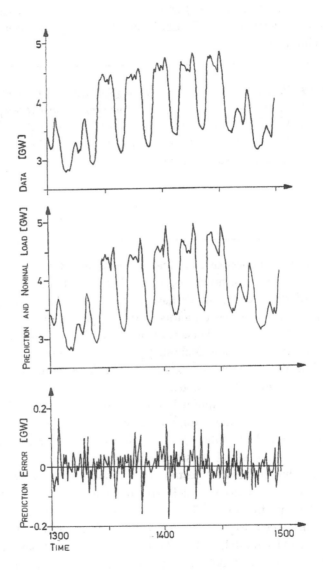

Figure 7.10 *Measured electricity consumption during one autumn week (top diagram), ARMA-predicted consumption, including weekday and hour correction, one hour ahead (middle diagram), and the prediction error (lower diagram).*

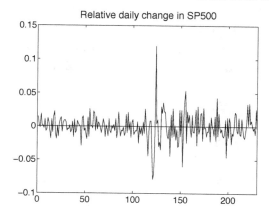

Figure 7.11 *Relative daily change,* $\log y_t - \log y_{t-1}$, *in the S&P500 stock market index for 230 days in 2002–2003.*

7.5   A simple non-linear model – the GARCH-process

As mentioned in Section 4.5.4, Gaussian models specified by a covariance function or spectrum have their limitations: they are not good models for processes where there is an irregular change between periods with large and small variability. In particular, in finance one uses the term *volatility* for the expected variability of the value of an economic asset in the near future, and it is common that periods with low and high volatility follow one another in an irregular way.

Figure 7.11 illustrates the daily relative changes, $x_t = \log(y_t/y_{t-1}) = \log y_t - \log y_{t-1}$, for Standard & Poor's 500 stock market index for 230 days in 2002–2003. An ARMA-model, even of high order, would not catch the occasional outbreaks of high variability in the series.

The acronym GARCH stands for *generalized autoregressive conditional heteroscedasticity.*[2] These models describe the simultaneous time development of a time series together with its local variability, which is expressed as a function of past values, [9, 20]. The structure is similar to the ARMA-model, with the difference that the variability of future values is conditional of current and old values.

In the definition we let $\{Z_t, t \in \mathbb{Z}\}$ be a sequence of independent random variables with mean zero and variance one. Further, X_t denotes the variable of interest and σ_t represents the variability at time t, which is also a stochastic process.

---

[2]A model with non-constant variance is called heteroscedastic.

Definition 7.3. *The bivariate process* $\{X_t, \sigma_t\}$ *is called a GARCH(p,q)-process with volatility process* $\{\sigma_t\}$ *if there exist constants* $\alpha_0 > 0$, $\alpha_1, \ldots, \alpha_p \geq 0$, $\beta_1, \ldots, \beta_q \geq 0$, *such that*

$$X_t = \sqrt{\sigma_t^2 Z_t}, \tag{7.24}$$

$$\sigma_t^2 = \alpha_0 + \sum_1^p \alpha_k X_{t-k}^2 + \sum_1^q \beta_k \sigma_{t-k}^2. \tag{7.25}$$

To specify the correct order, it is assumed that $\alpha_p > 0, \beta_q > 0$.

We give some basic facts about the GARCH(p,q)-model; see [10] or [47].

Stationarity condition: In order that there is a weakly stationary process that solves the GARCH-equations it is necessary and sufficient that $\sum_1^p \alpha_k + \sum_1^q \beta_k < 1$.

Mean and variance: The stationary GARCH(p,q)-process has mean zero and variance

$$\mathsf{E}[\sigma_t^2] = \frac{\alpha_0}{1 - \sum_1^p \alpha_k - \sum_1^q \beta_k}.$$

Covariance structure: The GARCH(p,q)-process $\{X_t\}$ is white noise, i.e., $\mathsf{C}[X_t, X_{t+\tau}] = 0$.

Remark 7.3. *The GARCH-model is an example of a model with parametric excitation. Inserting* $X_t^2 = \sigma_t^2 Z_t^2$ *into (7.25) and rearranging, we get (if* $p = q$*)*

$$\sigma_t^2 = \alpha_0 + \sum_1^p \left(\beta_k + \alpha_k Z_{t-k}^2 \right) \sigma_{t-k}^2,$$

which is an AR(p)-process with variable parameters and an extra off-set term. We note that the equation lacks the additive innovation term. The excitation instead shows up in the AR-parameters.

Example 7.7 ("GARCH(1,1)-process"). The GARCH(1,1)-model has been successfully used to describe many economic time series, and it is often sufficient for practical purposes. Its structure is simple; expressed in terms of the

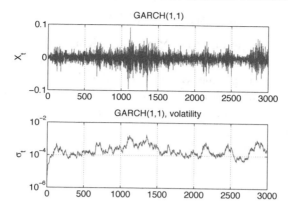

Figure 7.12 *Simulated GARCH(1,1)-model for the Swedish OMXS 30-index.*

squared series the model can be written as

$$X_t^2 = \sigma_t^2 Z_t^2,$$
$$\sigma_t^2 = \alpha_0 + \alpha_1 X_{t-1}^2 + \beta_1 \sigma_{t-1}^2 = \alpha_0 + (\beta_1 + \alpha_1 Z_{t-1}^2)\sigma_{t-1}^2.$$

Figure 7.12 shows a simulation of a GARCH(1,1)-model with parameters $\alpha_0 = 0.0000019$, $\alpha_1 = 0.0775$, $\beta_1 = 0.9152$ estimated from the Swedish OMXS 30-index. The upper plot shows the log-differences $X_t = \log Y_t / Y_{t-1}$ and the lower plot shows the volatility σ_t. ▲

The GARCH(p,q)-model generates uncorrelated X_t-variables, and hence 0 is the best (mean square) predictor of future values. Thus, if a time series can be well modeled by a GARCH process, this implies that prediction of future values is not possible. This, in fact, based both on extensive practical experience and economic theory, is known to be the case for many financial time series. Instead GARCH models offer a useful way to predict future volatility, and hence to measure investment risks, and it also provides a good way to simulate the time series. Financial time series often seem to be exposed to "structural changes," and therefore GARCH models often are useful only for limited periods of time, after which they have to be re-estimated.

Extensions of the GARCH-model have been developed to take care of dynamics in the observation equation, by replacing X_t in (7.24) and (7.25) by $X_t + \mu_t$ where μ_t is an AR-process. Other modifications include asymmetric excitations, external chocks, etc., in order to make the model more realistic; see [47], which also contains many other examples of non-linear time series.

7.6   Monte Carlo simulation of ARMA processes

Monte Carlo simulation of AR-, MA-, and ARMA-processes is simple. One technique is to apply a digital filter on a sequence of independent variables $(e_1,\ldots,e_n)$, for example the MATLAB-command x = filter(C,A,e). After a period of stabilization, the x-sequence is (approximately) an ARMA-process.

To get an exact Gaussian ARMA-process one can generate a Gaussian starting vector $(x_1,\ldots,x_{\max(p,q)})$ with the correct covariances, and then use the recursive ARMA-structure. The covariances have to be found from (7.22).

Exercises

7:1. Suppose that an AR(1)-process, $\{X_t, t = 0, \pm 1, \ldots\}$, has the covariance function $r(k) = 10 \cdot 0.5^{|k|}$. Which AR(1)-process is it? What is the variance of the innovations?

7:2. Find the spectral density, $R(f)$, and the covariance function, $r(k)$, at time lag zero, one and two, of the AR(2)-process

$$X_t - X_{t-1} + 0.5X_{t-2} = e_t,$$

where $\{e_t\}$ is a white innovation process with variance $V[e_t] = 4$.

7:3. The covariance function of a signal, $\{Y_t\}$, in discrete time, has been well estimated at a few time lags:

τ:	0	1	2	3	4
$r_Y(\tau)$:	2.4	1.6		-0.4	-0.6

We decide to use the following AR(2)-process to model the signal,

$$Y_t - Y_{t-1} + a_2 Y_{t-2} = e_t.$$

Find the innovation variance, the second AR-parameter, and the covariance function at time lag two.

7:4. The MA(q)-process

$$Y_t = e_t + b_1 e_{t-1} + \cdots + b_q e_{t-q},$$

where $\{e_t\}$ is a white innovation process with unit variance, has covariance function $r(0) = \frac{5}{4}$, $r(1) = 0$, $r(2) = \frac{1}{2}$, and $r(k) = 0$, $|k| \geq 3$. Determine the order of the process, the MA-coefficients, and the spectral density.

7:5. Let $e_t, t \in \mathbb{Z}$, be independent random variables with expectation m and variance σ^2. Define, for a constant M,

$$X_t = e_t + 2e_{t-1} - e_{t-3}, \quad Z_t = X_t - M.$$

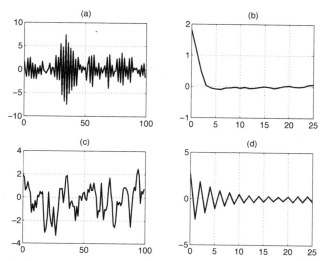

Figure 7.13 *Realizations and estimated covariance functions of one AR(1)- and one MA(2)-process in Exercise 7:6.*

(a) Choose M, such that $\{Z_t\}$ is an MA(3)-process. Determine the expectation, covariance function, and spectral density of $\{Z_t\}$.

(b) Find expectation, covariance function, and spectral density of $\{X_t\}$.

7:6. Let $\{X_t, t = 0, \pm 1, \dots\}$ be a stationary AR(1)-process,

$$X_t + 0.8X_{t-1} = e_t,$$

and let $\{Y_t, t = 0, \pm 1, \dots\}$ be an MA(2)-process,

$$Y_t = u_t + 0.8u_{t-1} + 0.4u_{t-2},$$

where $\{e_t\}$ and $\{u_t\}$ are independent white Gaussian processes with unit variance. Realizations and estimated covariance functions of $\{X_t\}$ and $\{Y_t\}$ are shown in Figure 7.13, but which subplot shows what?

7:7. (a) Combine, in Figure 7.14, realizations, covariance functions, and spectral densities for two stationary processes.

(b) Combine the processes with the correct pole-zero plot in Figure 7.15, where 'x' depicts poles and 'o' zeros. The processes are of the type AR or MA. State the type and order of the processes.

7:8. Figure 7.16 shows pole-zero plots, spectral densities, and covariance functions for three AR(2)-processes. Determine which figures that describe the same process.

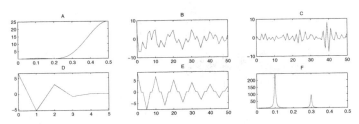

Figure 7.14 *Realizations, covariance functions, and spectral densities to Exercise 7:7a.*

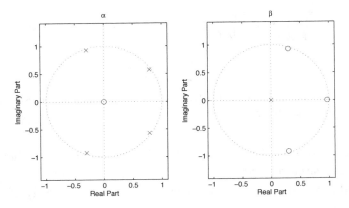

Figure 7.15 *Pole-zero plots to Exercise 7:7b.*

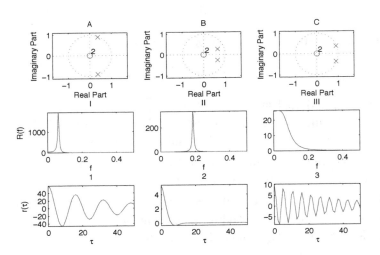

Figure 7.16 *Pole-zero plots, covariance functions, and spectral densities of two processes in Exercise 7:8.*

7:9. (This is a classical example of time series analysis in hydrology.) The Wabash river water level, W_t (meter), is measured once every month. The following AR-type model has been suggested:

$$W_t - 1.559\,W_{t-1} + 0.81\,W_{t-2} = 3.5 + e_t,$$

where $\{e_t\}$ is a white noise process with variance 4. What is the expected water level according to this model? What is the standard deviation?

7:10. Let $\{X_k, k \in \mathbb{Z}\}$ be an AR(1)-process, $X_k + a_1 X_{k-1} = e_k$, with $V[e_k] = \sigma^2$. Construct a new process, $\{Y_m, m \in \mathbb{Z}\}$, by using every second element of $\{X_k\}$, that is: $Y_m = X_{2m}$. Prove that $\{Y_m\}$ is an AR(1)-process and that it thus can be written in the following form:

$$Y_m + aY_{m-1} = u_m.$$

Also, compute a and $V[u_m]$. Hint: First, discuss what is needed for a process to be an AR(1)-process. Convince yourself that the process is of the form $Y_m + aY_{m-1} = u_m$, and that you need to prove that u_m and u_n, $m \neq n$, are independent and that Y_m and u_n are independent for all $m < n$.

7:11. Compute the cross-covariance, $C[X_{t+k}, e_t]$, $k = \ldots, -1, 0, 1, \ldots$, where $\{X_t\}$ is an AR(1)-process: $X_t = 0.4X_{t-1} + e_t$, and $V[e_t] = 2$.

7:12. A random process, $\{X_t\}$, is generated by filtering white noise:

$$X_t + a_1 X_{t-1} = e_t, \quad E[e_t] = 0, \quad V[e_t] = \sigma^2,$$

where a_1 and σ^2 are unknown. A simulation gives the following values:

$x_1, \ldots, x_8$	-0.15	-0.97	1.35	0.22	-0.79	-0.52	-0.57	-0.92
$x_9, \ldots, x_{15}$	-0.20	1.35	-1.32	-0.27	0.02	0.59	-0.83	

(a) Estimate $r_X(0)$ and $r_X(1)$, using the given observations.

(b) Use your estimates in (a) to estimate a_1 and σ^2.

7:13. Let $\{X_t\}$ be stationary with mean value m and assume

$$(X_t - m) + 0.25(X_{t-1} - m) = e_t,$$

where $\{e_t\}$ is Gaussian white noise with mean value zero and unit variance. The following values of X_t are observed:

$x_1, \ldots, x_5$	0.531	0.944	1.693	-0.390	0.171
$x_6, \ldots, x_{10}$	1.126	-0.497	0.431	2.145	1.193

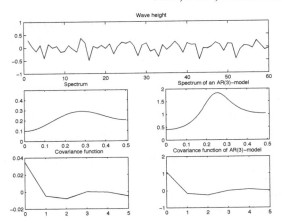

Figure 7.17 *Wave height, spectrum, and covariance function estimated using data, and spectrum and covariance function of the model in Exercise 7:15.*

Estimate m and compute a 95% confidence interval for your estimate.

7:14. Find the best one-step ahead predictor of the ARMA(1,1)-process defined by $X_t - 0.5X_{t-1} = e_t + 0.5e_{t+1}$.

7:15. Figure 7.17 shows the mean wave height, X_t, outside the Pakistan coast, as measured by a satellite passing the Indian Ocean on May 21, 1993. The process is well described by an AR(3)-model:

$$X_t + 0.2X_{t-1} + 0.27X_{t-2} + 0.0824X_{t-3} = \sigma e_t,$$

where e_t is white noise with unit variance. Compute $r_X(4)$ and $r_X(5)$ from the first four covariances:

τ:	0	1	2	3
$r_X(\tau)$:	0.3326	-0.0474	-0.0764	0.0007

Chapter 8

Linear filters – applications

8.1 Introduction

The AR, MA, and ARMA models in Chapter 7 form a very flexible class of stationary time series models. All their statistical properties can be derived from the covariance functions and the spectral formulation of input/output relations in linear filters. In this chapter we have collected some further applications in different fields of these relations.

The chapter starts with the continuous time analogue to the AR(p)-process, namely linear stochastic differential equations, with constant coefficients, driven by a stochastic process. We present spectral relations between the driving process and the response process, and in particular treat the case when the input is white noise.

Section 8.3 deals with the envelope of a stationary process used in many engineering applications to describe slowly varying amplitudes. It is here defined by means of the Hilbert transform.

In Sections 8.4 and 8.5, we treat two important applications in signal processing, the matched filter, used to detect the presence or absence of a characteristic signal pattern in a noisy environment, and the Wiener filter, used to separate disturbing noise, for example from an interesting audio or video signal. The principles behind the important Kalman filter is introduced in Section 8.6 as a recursive alternative for reconstruction of noisy data.

Finally, in Section 8.7, we describe how a linear filter and stationary processes can be used to optimize the performance of a mechanical system, namely the suspension of a car.

8.2 Differential equations with random input

8.2.1 Linear differential equations with random input

A linear stochastic differential equation is the continuous time analogue of the discrete time AR(p)-process. We start with a first order equation. Suppose

$\{X(t), t \in \mathbb{R}\}$ is a continuous stationary process with mean 0 and spectral density $R_X(f)$. Then, for constants a_0, a_1 with $\alpha = a_1/a_0 > 0$, the stochastic process $\{Y(t), t \in \mathbb{R}\}$, defined by

$$Y(t) = \frac{1}{a_0} \int_{-\infty}^{t} e^{-\alpha(t-u)} X(u) \, du, \tag{8.1}$$

is also stationary. In Example 6.2, we found its spectral density to be

$$R_Y(f) = \frac{1}{a_1^2 + (2\pi f a_0)^2} R_X(f). \tag{8.2}$$

Now, we recognize (8.1) as the solution to a stochastic version of the ordinary differential equation

$$a_0 Y'(t) + a_1 Y(t) = X(t). \tag{8.3}$$

Thus, we have formulated the solution to (8.3) as the output of a linear filter (8.1) with impulse response $h(u) = e^{-\alpha u}/a_0$, $u > 0$, and we can make the important observation that also $h(u)$ satisfies a differential equation,

$$a_0 h'(u) + a_1 h(u) = 0,$$

but now with vanishing right hand side and initial condition $h(0) = 1/a_0$.

We next consider a linear differential equation of order p, with constant coefficients,

$$a_0 Y^{(p)}(t) + a_1 Y^{(p-1)}(t) + \ldots + a_{p-1} Y'(t) + a_p Y(t) = X(t). \tag{8.4}$$

To formulate the solution and the statistical properties of the solution, we define the *generating polynomial*,

$$A(r) = a_0 + a_1 r + \ldots + a_p r^p,$$

and the *characteristic equation*,

$$r^p A(r^{-1}) = a_0 r^p + a_1 r^{p-1} + \ldots + a_{p-1} r + a_p = 0. \tag{8.5}$$

The differential equation (8.4) is called *stable* if the roots of the characteristic equation all have negative real part.

Theorem 8.1. *If the equation (8.4) is stable, i.e., the roots of the characteristic equation all have negative real part, and if the right hand side $\{X(t), t \in \mathbb{R}\}$ is a stationary stochastic process with spectral density $R_X(f)$, then there exists a stationary process $\{Y(t), t \in \mathbb{R}\}$ that solves the equation. It can be expressed as the output of a linear causal filter (i.e., $h(u) = 0$ for $u > 0$)*

$$Y(t) = \int_{-\infty}^{t} h(t-u)X(u)\,du, \tag{8.6}$$

whose impulse response function $h(u)$ is the solution to

$$a_0 h^{(p)}(t) + a_1 h^{(p-1)}(t) + \ldots + a_{p-1} h'(t) + a_p h(t) = 0,$$

with the initial conditions $h(0) = h'(0) = \ldots = h^{(p-2)}(0) = 0$, $h^{(p-1)}(0) = 1/a_0$. Further, $\int_0^\infty |h(u)|\,du < \infty$. The spectral density is

$$R_Y(f) = \frac{R_X(f)}{|a_p + a_{p-1}(i2\pi f) + \ldots + a_1(i2\pi f)^{p-1} + a_0(i2\pi f)^p|^2}.$$

Example 8.1. We want to solve the equation

$$Y''(t) + 2Y'(t) + Y(t) = X(t),$$

when $\{X(t), t \in \mathbb{R}\}$ is a stationary process. The impulse response $h(u)$ in (8.6) is a solution to

$$h''(t) + 2h'(t) + h(t) = 0,$$

and has the general form

$$h(t) = e^{-t}(C_1 + C_2 t),$$

with initial conditions giving $C_1 = 0, C_2 = 1$. The solution,

$$Y(t) = \int_{-\infty}^{t} (t-u)e^{-(t-u)}X(u)\,du,$$

has spectral density

$$R_Y(f) = \frac{R_X(f)}{|1 + 2(i2\pi f) + (i2\pi f)^2|^2} = \frac{R_X(f)}{(1 + (2\pi f)^2)^2}. \qquad \blacktriangle$$

Example 8.2 ("RC-filter"). In an RC-filter with input potential $X(t)$, the output potential follows the equation $RCY'(t) + Y(t) = X(t)$, with solution

$$Y(t) = \frac{1}{RC} \int_{-\infty}^{t} e^{-(t-u)/(RC)} X(u) \, du.$$

The frequency response was derived in Example 6.2, directly from the definition. We now illustrate a shortcut via a very simple formal scheme. Consider the differential equation and the implied equation for the frequency response:

$$RCY'(t) \qquad + \; Y(t) \; = \; X(t),$$
$$\downarrow \qquad\qquad \downarrow \qquad \downarrow$$
$$RC(i2\pi f)H(f) + H(f) = \; 1,$$

where each element in the second row consists of the frequency function from $\{X(t)\}$ to the corresponding term in the differential equation: the frequency response functions from $X(t)$ to $X(t)$ is identically 1, that from $X(t)$ to $Y(t)$ is $H(f)$, and that from $X(t)$ to $RCY'(t)$ goes first to $Y(t)$ with response function $H(f)$, and then further on to $RCY'(t)$ with response function $RC(i2\pi f)$.

Solving for $H(f)$, we get again,

$$H(f) = \frac{1}{1 + i2\pi fRC},$$

and, as a special case of (8.2),

$$R_Y(f) = |H(f)|^2 R_X(f) = \frac{R_X(f)}{1 + (2\pi fRC)^2}. \tag{8.7}$$

We could reach the same results by a much lengthier argument, via the covariance function relation between $\{X(t)\}$ and $\{Y(t)\}$:

$$r_X(\tau) = C[RCY'(t) + Y(t), RCY'(t+\tau) + Y(t+\tau)]$$
$$= (RC)^2 r_{Y'}(\tau) + RC r_{Y,Y'}(t, t+\tau) + RC r_{Y',Y}(t, t+\tau) + r_Y(\tau)$$
$$= (RC)^2 r_{Y'}(\tau) + RC r'_Y(\tau) + RC r'_Y(-\tau) + r_Y(\tau)$$
$$= (RC)^2 r_{Y'}(\tau) + RC r'_Y(\tau) - RC r'_Y(\tau) + r_Y(\tau) = (RC)^2 r_{Y'}(\tau) + r_Y(\tau),$$

according to Theorem 6.4. Then Theorem 6.3 gives

$$r_X(\tau) = (RC)^2 \int e^{i2\pi f\tau} (2\pi f)^2 R_Y(f) \, df + \int e^{i2\pi f\tau} R_Y(f) \, df$$
$$= \int e^{i2\pi f\tau} \{ (2\pi fRC)^2 + 1 \} R_Y(f) \, df,$$

and we can identify $R_X(f) = \{ (2\pi fRC)^2 + 1 \} R_Y(f)$ as the spectral density of $\{X(t)\}$, which is equivalent with (8.7). ▲

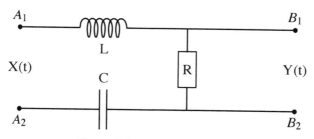

Figure 8.1 *Resonance circuit.*

Example 8.3 ("Resonance circuit"). An electrical resonance circuit consists of an inductance, a resistance, and a capacitance connected as in Figure 8.1. It acts as a bandpass filter. The input potential $X(t)$ between A_1 and A_2 and the current $I(t)$ in the circuit obey the following equation,

$$LI'(t) + RI(t) + \frac{1}{C}\int_{-\infty}^{t} I(s)\,ds = X(t).$$

The potential $Y(t) = RI(t)$ over the resistor between B_1 and B_2 and the frequency response function $H(f)$ for the filter therefore satisfy the equations

$$
\begin{array}{ccccccc}
LY''(t) & + & RY'(t) & + & \frac{1}{C}Y(t) & = & RX'(t), \\
\downarrow & & \downarrow & & \downarrow & & \downarrow
\end{array}
\qquad (8.8)
$$

$$L(i2\pi f)^2 H(f) + R(i2\pi f)H(f) + \frac{1}{C}H(f) = R(i2\pi f),$$

leading to

$$H(f) = \frac{i2\pi f R/L}{-(2\pi f)^2 + i2\pi f R/L + 1/(LC)}.$$

The resonance frequency f_0 and relative bandwidth $1/Q$ are defined as

$$2\pi f_0 = 1/\sqrt{LC}, \quad \text{and} \quad 1/Q = \Delta f/f_0 = R\sqrt{C/L},$$

where $\Delta f = f_2 - f_1$ is determined by $|H(f_1)| = |H(f_2)| = |H(f_0)|/\sqrt{2}$. Figure 8.2 illustrates how the bandwidth affects the frequency content after the filter. ▲

8.2.2 *Differential equations driven by white noise*

The important autoregressive discrete time AR(p)-process in Section 7.2.1 is driven by a white noise sequence, e_t, which does not depend on the process history before time t. Similarly, one would like to use continuous time

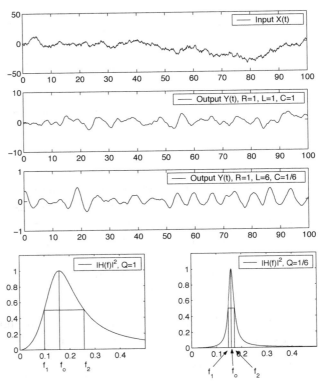

Figure 8.2 *Bandpass filter with different bandwidth: broadband with $R = L = 1$ and $Q = 1$, and narrowband with $R = 1, L = 6, C = 1/6$, and $Q = 1/6$. In both cases $f_0 = 1/(2\pi)$.*

white noise as the input variable $X(t)$ in the right hand side of the differential equation (8.4). However, the literal translation of discrete time white noise to continuous time, as a process where the values at different points of time are independent, is not useful. A possibility, as discussed in Chapter 4, could be to instead consider stochastic processes whose realizations are generalized functions. However, this is only useful in linear problems.

The more useful alternative is to integrate the differential equation and use a Wiener process as the right hand side of this equation. This idea has been very extensively generalized, and has developed into an entire large area of probability theory called *stochastic calculus*, with many important applications. In this section we give a very small preview of this area, and in particular give a first definition of the so-called Ito stochastic integral. For more information we refer the interested reader to the book [60].

Example 8.4 ("Brownian motion and velocity"). In the Brownian motion example in Section 5.3 we described the *location* of a particle suspended in a fluid as a Wiener process. Suppose now that we instead want to describe the velocity, where we for simplicity only consider movements along the real line only. Then the velocity $Y(t)$ of the particle at time t, exposed to a varying outer force $X(t)$, obeys the differential equation

$$a_0 Y'(t) + a_1 Y(t) = X(t), \qquad (8.9)$$

with the solution

$$Y(t) = \frac{1}{a_0} \int_{-\infty}^{t} e^{-a_1(t-u)/a_0} X(u) \, du. \qquad (8.10)$$

Here, a_1 depends on the viscosity of the fluid, and a_0 is the particle mass.

If the force $X(t)$ is the result of collisions between the particle and the molecules in the fluid, it would be natural to think that different $X(t)$-values are random and virtually independent, i.e., that they are white noise in continuous time, since there would be of the order 10^{21} collisions per second; [13].

However, what is then white noise in continuous time? ▲

The solution is to integrate (8.9) and let the integrated right hand side of the equation be the increments of a Wiener process

$$\int_{a}^{b} X(t) \, dt = W(b) - W(a) = \int_{a}^{b} W'(t) \, dt = \int_{a}^{b} dW(t).$$

Here, the last equality above is just a question of notation: in stochastic calculus one always writes $dW(t)$ instead of $W'(t)dt$. The reason of course is that since $W(t)$ is not differentiable (Example 6.7), it is not clear what $W'(t)$ should mean.

Now, to be able to integrate with respect to $dW(u)$ we require a new definition of an integral, as a stochastic Stieltjes integral (cf. Appendix B.6), defined as a limit,

$$\int_{u=a}^{b} g(u) \, dW(u) = \lim_{n \to \infty} \sum_{k=0}^{n-1} g(u_k)(W(u_{k+1}) - W(u_k)),$$

when $a = u_0 < u_1 < \ldots < u_n = b$ is a successively finer and finer subdivision of the interval $[a, b]$. The limit exists if $g(u)$ is of "bounded variation." More restrictively, the limit exists if $g(t)$ is differentiable, and the integral can then be evaluated by partial integration,

$$\int_{u=a}^{b} g(u) \, dW(u) = g(b)W(b) - g(a)W(a) - \int_{u=a}^{b} g'(u)W(u) \, du. \qquad (8.11)$$

Example 8.5 ("Ornstein-Uhlenbeck process"). With this new integral definition equations of the type $Y'(t) + \alpha Y(t) = \sigma_0 W'(t)$ are written as

$$dY(t) + \alpha Y(t)dt = \sigma_0 dW(t), \text{ with } Y(t) = \sigma_0 \int_{-\infty}^{t} e^{-\alpha(t-u)} dW(u), \quad (8.12)$$

and have a mathematically rigorous interpretation as a shorthand notation for
the integral equation

$$Y(t) - Y(t_0) + \alpha \int_{t_0}^{t} Y(s)ds = \sigma_0 (W(t) - W(t_0)). \quad (8.13)$$

We have argued in Section 6.4 that white noise in continuous time should
have formally constant spectral density. In fact the solution in (8.12) can be
show to be an Ornstein-Uhlenbeck process, and hence, see Example 4.2, $Y(t)$
has spectral density

$$R_Y(f) = \frac{\sigma_0^2}{\alpha^2 + (2\pi f)^2} = \sigma^2 \frac{2\alpha}{\alpha^2 + (2\pi f)^2}.$$

This agrees with (8.1) and (8.2) if we take $R_X(f) = \sigma^2 = \sigma_0^2/(2\alpha)$.

The following consequence of (8.12) is often useful:

$$Y(t + \tau) = \sigma_0 \int_{-\infty}^{t} e^{-\alpha(t+\tau-u)} dW(u) + \sigma_0 \int_{t}^{t+\tau} e^{-\alpha(t+\tau-u)} dW(u)$$

$$= Y(t)e^{-\alpha\tau} + \sigma_0 \int_{t}^{t+\tau} e^{-\alpha(t+\tau-u)} dW(u). \quad (8.14)$$

The increment integral has a normal distribution and is independent of $Y(t)$
and one can see the AR(1) structure of the process. The increment variance is

$$V\left[\sigma_0 \int_{t}^{t+\tau} e^{-\alpha(t+\tau-u)} dW(u)\right] = \sigma_0^2 \int_{0}^{\tau} e^{-2\alpha(\tau-u)} du = \sigma^2(1 - e^{-2\alpha\tau}).$$

$$(8.15)$$

▲

A solution to the general p^{th} order equation,

$$a_0 Y^{(p)}(t) + a_1 Y^{(p-1)}(t) + \ldots + a_{p-1} Y'(t) + a_p Y(t) = \sigma_0 W'(t),$$

can be obtained in a similar way as $Y(t) = \sigma_0 \int_{-\infty}^{t} h(t - u) dW(u)$, with the
same impulse response function $h(u)$ as in Theorem 8.1. The spectral density
will be

$$R_Y(f) = \frac{\sigma_0^2}{|a_p + a_{p-1}(i2\pi f) + \ldots + a_1(i2\pi f)^{p-1} + a_0(i2\pi f)^p|^2},$$

and the formal differential equation is replaced by a well-defined differential-integral equation,

$$a_0(Y^{(p-1)}(t) - Y^{(p-1)}(t_0)) + a_1(Y^{(p-2)}(t) - Y^{(p-2)}(t_0))$$
$$+ \ldots + a_{p-1}(Y(t) - Y(t_0)) + a_p \int_{t_0}^{t} Y(u)\, du = \sigma_0(W(t) - W(t_0)).$$

Remark 8.1. *As above, stochastic differential equations with white noise are usually written in differential form, as*

$$a_0 dY(t) + a_1 Y(t)\, dt = \sigma\, dW(t),$$

or, more generally,

$$dY(t) = a(t)Y(t)\, dt + \sigma(t)\, dW(t),$$

with variable, but still deterministic coefficients.
    For even more generality, one can let the coefficients be stochastic, as in

$$dY(t) = a(Y(t), t)Y(t)\, dt + \sigma(Y(t), t)\, dW(t).$$

This poses new and more difficult problems, and we leave these to specialized courses in stochastic calculus; [60].

8.3   The envelope

An envelope wraps up and contains something. A mathematical *envelope* to a rapidly oscillating function $X(t)$ is a slowly varying function $Z(t)$ such that $|X(t)| \leq Z(t)$ for all t, and $Z(t)$ and $X(t)$ come as close as its smoothness permits. The curves $\pm Z(t)$ enclose the curve $X(t)$ and are a tangent for some t-values. The most well known example of an envelope is AM radio (Amplitude Modulated radio transmission), in which the audio signal is the envelope of a rapidly oscillating carrier wave, with frequency equal to the frequency of the radio station.

Example 8.6 ("Amplitude modulated cosine function"). The function $X(t)$, defined by

$$X(t) = (2 + \cos 2\pi t)\cos 20\pi t,$$

consists of a rapid oscillation with frequency 10 Hz, with a periodically changing amplitude, with period one second; see Figure 8.3. It can be expressed in an equivalent way as the sum of three cosine functions with almost the same frequency, $X(t) = 2\cos 20\pi t + \frac{1}{2}(\cos 18\pi t + \cos 22\pi t)$. The function $Z(t) = 2 + \cos 2\pi t$ is an envelope to $X(t)$.   ▲

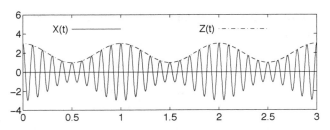

Figure 8.3 *The function* $X(t) = (2 + \cos 2\pi t)\cos 20\pi t$ *together with its envelope* $Z(t) = 2 + \cos 2\pi t$.

There is no unique definition of an envelope, and many functions can act in the role. We will use a definition that lends itself to mathematical analysis, and can be used both for deterministic and stochastic functions.

8.3.1   The Hilbert transform

First we define the Hilbert transform,[1] in terms of a linear time-invariant non-causal filter, defined by its frequency function $H(f)$.

Definition 8.1. *The Hilbert transform is the output of a linear filter working on* $\{X(t)\}$, *and with frequency function*

$$H(f) = \begin{cases} i & \text{for} \quad f < 0, \\ 0 & \text{for} \quad f = 0, \\ -i & \text{for} \quad f > 0. \end{cases}$$

It is a phase shifting filter that subtracts $\pi/2$ *from the phase if* $f > 0$, *and adds* $\pi/2$ *if* $f < 0$.

The corresponding impulse response function is $h(u) = 1/(\pi u)$ *and the filter integral form* $Y(t) = \frac{1}{\pi}\int_{-\infty}^{\infty} X(u)/(t-u)\,du$ *has to be interpreted as a Cauchy principal value.*

Since the Hilbert transform is a linear, time-invariant filter, a stationary process has a stationary Hilbert transform. In fact, the Hilbert transform will have almost identical spectral properties as the process itself, the only difference being the absence of a possible spectral component at the zero frequency.

---

[1] David Hilbert, German mathematician, 1862–1943.

Theorem 8.2. *Let $\{X(t)\}$ be a stationary stochastic process with mean value 0, covariance function $r_X(\tau)$, and spectral density $R_X(f) = b_0 \delta_0(f) + R_X^0(f)$ (so $R_X(f)$ may contain a delta function at the origin). Then its Hilbert transform $\{Y(t)\}$ is a stationary process, with the following properties.*

a) *$\{Y(t)\}$ has spectral density $R_Y(f) = R_X^0(f)$, and covariance function $r_Y(\tau) = r_X(\tau) - b_0$. If $b_0 = 0$, then $\{X(t)\}$ and $\{Y(t)\}$ have the same covariance function and spectral density functions.*

b) *The cross-covariance between $\{X(t)\}$ and $\{Y(t)\}$ is*

$$r_{XY}(\tau) = C[X(t), Y(t+\tau)] = 2 \int_0^\infty \sin(2\pi f \tau) R_X(f)\, df.$$

In particular, $X(t)$ and $Y(t)$ are uncorrelated; $r_{XY}(0) = C[X(t), Y(t)] = 0$. Note, that this does not mean that the processes $\{X(t)\}$ and $\{Y(t)\}$ are uncorrelated, since usually $C[X(s), Y(t)] \neq 0$ for $s \neq t$.

c) *If $\{X(t)\}$ is a Gaussian process, then so is $\{Y(t)\}$, and they have the same finite-dimension distributions if $b_0 = 0$.*

Proof. Part (a) follows directly from Theorem 6.2. Since $|H(f)| = 1$ for $f \neq 0$, the Hilbert transform $\{Y(t)\}$ has the same spectral density as $\{X(t)\}$, except a possible delta function at the origin, where $H(0) = 0$.

Part (b) follows from (6.20) in Theorem 6.7, which gives the cross-spectral density between $\{X(t)\}$ and $\{Y(t)\}$ as $R_{XY}(f) = H(f)R_X(f)$. Inserting $R_{XY}(f)$ in the expression for the cross-covariance function and using $(e^{i2\pi f\tau} - e^{-i2\pi f\tau})/(2i) = \sin 2\pi f\tau$, we get

$$r_{XY}(\tau) = \int_{-\infty}^\infty e^{i2\pi f\tau} R_{XY}(f)\, df = \int_{-\infty}^\infty e^{i2\pi f\tau} H(f)R_X(f)\, df$$

$$= \int_{-\infty}^{0^-} i e^{i2\pi f\tau} R_X(f)\, df + \int_{0^+}^\infty (-i) e^{i2\pi f\tau} R_X(f)\, df$$

$$= 2 \int_0^\infty \sin(2\pi f\tau) R_X(f)\, df.$$

Part (c), finally, is a consequence of the general property of a linear filter that it preserves Gaussianity. ∎

Example 8.7 ("Random phase and amplitude"). The Hilbert transform of

$$X(t) = A_0 + A\cos(2\pi f t + \phi)$$

is

$$Y(t) = A\cos(2\pi f t + \phi - \pi/2) = A\sin(2\pi f t + \phi),$$

since $\arg H(f) = -\pi/2$ for $f > 0$, and $\cos(x - \pi/2) = \sin x$. The Hilbert transform of $\cos t$ is $\sin t$, and that of $\sin t$ is $-\cos t$. Thus, we have the following pair of a process $X(t)$ and its Hilbert transform $Y(t)$,

$$X(t) = U_0 + \sum_{k=1}^{n} \{U_k \cos(2\pi f_k t) + V_k \sin(2\pi f_k t)\} = A_0 + \sum_{k=1}^{n} A_k \cos(2\pi f_k t + \phi_k),$$

$$Y(t) = \sum_{k=1}^{n} \{U_k \sin(2\pi f_k t) - V_k \cos(2\pi f_k t)\} = \sum_{k=1}^{n} A_k \sin(2\pi f_k t + \phi_k). \qquad \blacktriangle$$

8.3.2 A complex representation

The Hilbert transform $\{Y(t)\}$ of a stationary process $\{X(t)\}$ can be used to extract valuable information about the variability of $X(t)$ in the form of its envelope. We define the complex process $Z(t) = X(t) + iY(t)$. The following Thorem 8.3 is then a direct consequence of Theorem 8.2 and Example 5.6.

Definition 8.2. *The envelope of stationary process $\{X(t)\}$ with Hilbert transform $\{Y(t)\}$ is the absolute value*

$$Z_a(t) = |Z(t)| = \sqrt{X^2(t) + Y^2(t)}.$$

Theorem 8.3. *If $\{X(t)\}$ is a stationary Gaussian process without spectral mass at $f = 0$, and with $V[X(t)] = \sigma_X^2$, the envelope $Z_a(t)$ has a Rayleigh distribution with probability density*

$$f_{Z_a(t)}(x) = \frac{x}{\sigma_X^2} e^{-x^2/(2\sigma_X^2)}, \quad x > 0.$$

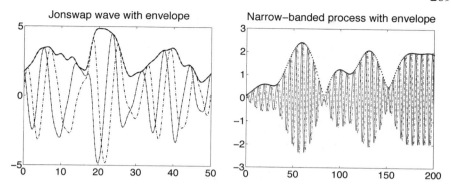

Figure 8.4 *Thin solid line: Gaussian process. Dash-dotted line: Hilbert transform of the process. Thick solid line: envelope of the process. Left: Jonswap broad-banded waves. Right: process with narrow rectangular spectrum.*

Example 8.8 ("Random phase and amplitude"). From Example 8.7 we get the complex representation

$$Z(t) = U_0 + \sum_{k=1}^{n} \left\{ U_k \big(\cos(2\pi f_k t) + i \sin(2\pi f_k t) \big) \right.$$
$$\left. - iV_k \big(\cos(2\pi f_k t) + i \sin(2\pi f_k t) \big) \right\} = X_0 + \sum_{k=1}^{n} (U_k - iV_k) e^{i2\pi f_k t}. \qquad \blacktriangle$$

8.3.3   The envelope of a narrow-banded process

For the envelope, it is obviously true that $Z_a(t) \geq |X(t)|$, with equality when $Y(t) = 0$. For a narrow-banded process much more can be said, and the interpretation of the envelope as the changing amplitude of a rapidly oscillating process gets a physical meaning. Figure 8.4 illustrates the behavior of the envelope for a broad-banded and a narrow-banded process.

Example 8.9. The simplest example of a narrow-banded process is the sum of two cosine functions with almost the same frequencies, $f_0 - \delta/2$ and $f_0 + \delta/2$, respectively,

$$X(t) = A_1 \cos(2\pi(f_0 - \delta/2)t + \phi_1) + A_2 \cos(2\pi(f_0 + \delta/2)t + \phi_2).$$

The Hilbert transform is

$$Y(t) = A_1 \sin(2\pi(f_0 - \delta/2)t + \phi_1) + A_2 \sin(2\pi(f_0 + \delta/2)t + \phi_2).$$

Since $\cos^2 x + \sin^2 x = 1$, one has, after some simplification,

$$X^2(t) + Y^2(2) = A_1^2 + A_2^2 + 2A_1 A_2 \cos(2\pi \delta t - \phi_1 + \phi_2).$$

The complex representation can be written in the form

$$Z(t) = X(t) + iY(t) = e^{i2\pi f_0 t}\{A_1 \cos(\pi\delta t - \phi_1) + A_2 \cos(\pi\delta t + \phi_2)$$
$$- i(A_1 \sin(\pi\delta t - \phi_1) - A_2 \sin(\pi\delta t + \phi_2))\},$$

showing a rapid oscillation, $e^{i2\pi f_0 t}$, with slowly varying amplitude.   ▲

Let us now consider the envelope of a narrow-banded Gaussian process with a spectral density symmetrically located around the frequencies $\pm f_0$,

$$R_X(f) = \frac{1}{2}R_0(f + f_0) + \frac{1}{2}R_0(f - f_0), \tag{8.16}$$

where $R_0(x)$ is an even function, which is 0 outside an interval $|x| \le d$, with $d < f_0$; see Figure 8.5.

Theorem 8.4. *Let $\{X(t)\}$ be a narrow-banded stationary Gaussian process with mean 0 and a spectral density of the form (8.16) (hence without delta component at $f = 0$), and let $\{Y(t)\}$ be its Hilbert transform. Then, there exists two independent, identically distributed Gaussian processes, $\{X_1(t)\}$ and $\{X_2(t)\}$, with spectral density $R_0(f)$, such that the complex representation $\{Z(t)\}$ can be written in the form*

$$Z(t) = X(t) + iY(t) = e^{i2\pi f_0 t}(X_1(t) + iX_2(t)). \tag{8.17}$$

The envelope $\{Z_a(t)\}$ can be represented by the independent $\{X_1(t)\}$ and $\{X_2(t)\}$,

$$Z_a(t) = \sqrt{X^2(t) + Y^2(t)} = \sqrt{X_1^2(t) + X_2^2(t)}. \tag{8.18}$$

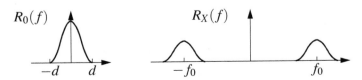

Figure 8.5 *Narrow-banded, symmetric spectrum.*

Proof. Multiplying in (8.17) by $e^{-i2\pi f_0 t}$ and taking the real and imaginary parts, we can identify $X_1(t)$ and $X_2(t)$ as

$$X_1(t) = \Re\left\{e^{-i2\pi f_0 t}(X(t) + iY(t))\right\} = X(t)\cos 2\pi f_0 t + Y(t)\sin 2\pi f_0 t,$$
$$X_2(t) = \Im\left\{e^{-i2\pi f_0 t}(X(t) + iY(t))\right\} = -X(t)\sin 2\pi f_0 t + Y(t)\cos 2\pi f_0 t.$$

Thus, both $X_1(t)$ and $X_2(t)$ are Gaussian processes, with mean 0, and they are also simultaneously Gaussian.

To prove the theorem we have to compute the covariance and cross-covariance functions, and this can be done by rather lengthy manipulations with trigonometric formulas. One finds

$$\mathsf{C}[X_1(t), X_1(t+\tau)]$$
$$= r_X(\tau)\cos(2\pi f_0 t)\cos(2\pi f_0(t+\tau)) + r_{XY}(\tau)\cos(2\pi f_0 t)\sin(2\pi f_0(t+\tau))$$
$$+ r_{XY}(-\tau)\sin(2\pi f_0 t)\cos(2\pi f_0(t+\tau)) + r_Y(\tau)\sin(2\pi f_0 t)\sin(2\pi f_0(t+\tau))$$
$$= r_X(\tau)\cos(2\pi f_0 \tau) + r_{XY}(\tau)\sin(2\pi f_0 \tau). \tag{8.19}$$

Calculations with $X_2(t)$ lead to the same result, and we have shown that $\{X_1(t)\}$ and $\{X_2(t)\}$ are stationary with a common covariance function, which we denote $r_0(\tau)$.

To show that the common spectral density is $R_0(f)$, we use the spectral representations of $r_X(\tau)$ and $r_{XY}(\tau)$ in (8.19). For $r_X(\tau)$, we use the real form, i.e., Equation (4.5), and for $r_{XY}(\tau)$ we use Theorem 8.2(b). The result is

$$r_0(\tau) = 2\int_0^\infty \cos(2\pi f\tau)\cos(2\pi f_0\tau)R_X(f)\,df$$
$$+ 2\int_0^\infty \sin(2\pi f\tau)\sin(2\pi f_0\tau)R_X(f)\,df$$
$$= 2\int_0^\infty \cos(2\pi(f-f_0)\tau)R_X(f)\,df$$
$$= \int_0^\infty \cos(2\pi(f-f_0)\tau)R_0(f-f_0)\,df = \int_{-\infty}^\infty e^{i2\pi f\tau}R_0(f)\,df. \tag{8.20}$$

To show the independence, we calculate the cross-covariance function:

$$\mathsf{C}[X_1(t), X_2(t+\tau)] = 2\int_0^\infty \cos\left(2\pi(f-f_0)\tau\right)R_X(f)\,df$$
$$= \int_0^\infty \sin\left(2\pi(f-f_0)\tau\right)R_0(f-f_0)\,df = 0, \tag{8.21}$$

for all τ, since $R_0(f)$ is symmetric. The processes are uncorrelated, and hence independent, since they are Gaussian. ∎

Theorem 8.4 can be used to model noise in FM radio communication. A radio signal is a modulated carrier radio wave with frequency f_0 (= carrier frequency). When the signal is transmitted it is corrupted by adding noise in the form of a stationary Gaussian process. In the receiver, the signal is demodulated to remove the carrier and finally low-pass filtered. What remains after the filter is a "signal" with spectral density $R_0(f)$ plus a filtered noise spectrum. The assumed symmetry of $R_0(f)$ guarantees that the covariance $r_0(\tau)$ is real, (8.20). In practice, asymmetric $R_0(f)$ may well appear, and therefore it is necessary to introduce complex valued signals in radio communication theory.

Remark 8.2. *For a narrow-banded process $\{X(t)\}$, one can construct a simple approximation to the envelope by means of the derivative $X'(t)$; remember that both Hilbert transformation and differentiation shifts the phase by an amount $\pi/2$. Consider a process of the form*

$$X(t) = A(t)\cos(2\pi f_0 t + \phi_0),$$

where $A(t) > 0$ is a slowly varying amplitude function with $A'(t) \approx 0$. Then,

$$X'(t) \approx -A(t) \cdot (2\pi f_0) \cdot \sin(2\pi f_0 t + \phi_0),$$

and

$$X(t)^2 = A(t)^2 \cos^2(2\pi f_0 t + \phi_0),$$
$$X'(t)^2 = A(t)^2 \sin^2(2\pi f_0 t + \phi_0) \cdot (2\pi f_0)^2.$$

To find the slowly varying amplitude, just take $A(t)^2 \approx X(t)^2 + \frac{X'(t)^2}{(2\pi f_0)^2}$, and use

$$A(t) \approx \sqrt{X(t)^2 + \frac{X'(t)^2}{(2\pi f_0)^2}} \tag{8.22}$$

as an approximation to the true envelope.

8.4   Matched filter

The matched filter was invented during World War II, as an efficient aid to detect characteristic signals in a radar echo.[2] Its important feature is that it gives an optimal way to determine whether an object is present in a noisy

---

[2]Dwight O. North, American engineer, 1909–1998, introduced it and also described many of the characteristics, including the statistical properties that we focus on here.

signal, or not. It may be a radar echo, a spy satellite image, the return from an OCR device, or other device, where the object has more or less known shape. The simplest example is a digital communication system, where "zeros" and "ones" are transmitted as characteristic pulses, and the receiver simply shall determine whether zero or one was transmitted. This is also the context we shall use to describe the matched filter and its basic properties.

8.4.1 Digital communication

In a digital communication system, a message is transmitted as a stream of zeros and ones. Each symbol is transmitted as a special signal during a short time interval with length T. We assume that a "one" is indicated by the transmission of a deterministic signal $s(t), 0 \leq t \leq T$, with a form known by the receiver, while a "zero" is marked by the absence of the signal. The message is transmitted through a noisy transmission channel that adds random noise $\{N(t)\}$ to the transmission. The noise can be regarded as a Gaussian stationary process, either pure white noise or colored noise with know or unknown spectral properties.

The disturbed signal that is received is either signal plus noise, $s(t)+N(t)$, or only noise, $N(t)$. The task is to find a good way of treating the received information, to be able to determine, with an as small probability of error as possible, whether the signal $s(t)$ is present or not. This is done by a linear filter, acting on the received signal in each time slot, and the output is used to decide on "signal $s(t)$ present" (= "one" sent) or "signal $s(t)$ absent" (= "zero" sent). By optimal choice of filter characteristics, this detection can be made highly reliable, presuming the noise is not too strong.

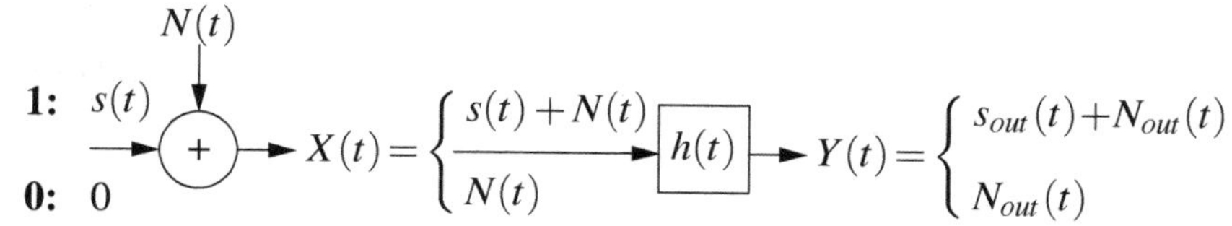

Figure 8.6 *Schematic illustration of matched filter to determine if $s(t)$ is present.*

The output from the filter is read off just at the end of each time slot, T, and used to distinguish between "one" and "zero." The output, $Y(T)$, will be different, depending of what was actually sent, as illustrated in Figure 8.6. The task is to decide which of the following two alternatives is at hand.

"One" sent: $s_{out}(T)+N_{out}(T) = \int_0^T h(T-u)s(u)\,du + \int_0^T h(T-u)N(u)\,du,$

"Zero" sent: $N_{out}(T) = \int_0^T h(T-u)N(u)\,du.$

Here, we have defined the deterministic $s_{out}(T) = \int_0^T h(T-u)s(u)\,du$ as the filtered signal, while the random $N_{out}(T) = \int_0^T h(T-u)N(u)\,du$ is the filtered noise.

8.4.2   A statistical decision problem

The problem to decide, based on the filter output $Y(T)$, whether it was a "one" or a "zero" that was sent, is a statistical decision problem. Formulated as a statistical hypothesis test, we have to decide which of two hypotheses,

$$H_0 : Y(T) = N_{out}(T) \qquad \text{("zero" sent)},$$
$$H_1 : Y(T) = s_{out}(T) + N_{out}(T) \quad \text{("one" sent)},$$

is the true one.

Now, the observed output $y(T)$ is just a real number, and the receiver has no chance to know if this number comes from the sum of $s_{out}(T)$ and $N_{out}(T)$, or just $N_{out}(T)$. However, $y(T)$ is an observation of the random variable $Y(T)$ and as such its *distribution* is different under the two alternatives.

Assuming the noise $N(t)$ is Gaussian with mean 0 and some covariance function $r_N(\tau)$, as it often is, then the filtered noise $N_{out}(T)$ has a normal distribution with mean 0 and variance,

$$\sigma_N(T)^2 = \mathsf{V}[N_{out}(T)] = \mathsf{E}[N_{out}(T)^2] = \int_0^T \int_0^T h(u)h(v)r_N(u-v)\,du\,dv,$$
(8.23)

according to Theorem 6.1. Thus, the hypotheses to be tested are

$$H_0 : Y(T) \sim N(0, \sigma_N^2),$$
$$H_1 : Y(T) \sim N(s_{out}(T), \sigma_N^2).$$

Let us assume $s_{out}(T) > 0$. Then it is reasonable to prefer hypothesis H_1 over H_0 if $y(T)$ is large, and the opposite if $y(T)$ is small. With a suitable *decision threshold k*, the test is

Test   Decision

$$y(T) \begin{cases} \leq k & \Rightarrow \text{ "zero" sent,} \\ > k & \Rightarrow \text{ "one" sent.} \end{cases}$$

The error probabilities are

$$\alpha = \mathsf{P}(\text{decide } H_1 \text{ when } H_0 \text{ is true}),$$
$$\beta = \mathsf{P}(\text{decide } H_0 \text{ when } H_1 \text{ is true}),$$

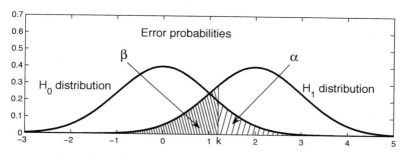

Figure 8.7 *Filter output distributions under two hypotheses.*

and these can be computed from the normal distribution; with Φ for standard normal distribution function,

$$\alpha = \mathsf{P}\left(Y(T) > k \mid Y(T) \sim N(0, \sigma_N^2)\right) = 1 - \Phi\left(\frac{k}{\sigma_N(T)}\right),$$

$$\beta = \mathsf{P}\left(Y(T) < k \mid Y(T) \sim N(s_{out}(T), \sigma_N^2)\right) = \Phi\left(\frac{k - s_{out}(T)}{\sigma_N(T)}\right)$$

$$= 1 - \Phi\left(\frac{s_{out}(T) - k}{\sigma_N(T)}\right).$$

By choosing an appropriate k-value, one can balance the two error probabilities according to what error is most serious. In digital transmission it may be natural to choose equal error probabilities, but in other detection situations, one type of error may be much more serious than the other, for example, in target detection and identification, when the question is to "pull the trigger" or not.

To get equal error probabilities one must take $k = s_{out}(T)/2$, which gives

$$\alpha = \beta = 1 - \Phi\left(\frac{s_{out}(T)}{2\sigma_N(T)}\right).$$

Obviously, the error probabilities become smaller with increasing ratio $s_{out}(T)/\sigma_N(T)$ or, equivalently, with increasing

$$s_{out}^2(T)/\sigma_N^2(T) = s_{out}^2(T)/\mathsf{E}[N_{out}(T)^2].$$

Definition 8.3. *In a detection problem, the ratio*

$$\text{SNR} = \frac{s_{out}^2(T)}{\mathsf{E}[N_{out}^2(T)]} = \frac{s_{out}^2(T)}{\sigma_N^2(T)}$$

is called the signal-to-noise ratio (SNR) at decision time T.

Thus, the signal-to-noise ratio in a decision problem is the square of the ratio between the filtered signal and the standard deviation of the filtered noise, at the decision time. Even if the noise is not Gaussian, the SNR is used as a measure of efficiency of a detection filter.

8.4.3   The optimal matched filter

Our next problem is to find the optimal detection filter, i.e., the impulse response $h(u)$ of the filter that maximizes the signal-to-noise ratio

$$\text{SNR} = \frac{s_{out}^2(T)}{\mathsf{E}[N_u^2(T)]} = \frac{\{\int s(T-u)h(u)\,du\}^2}{\int\int h(u)h(v)r(u-v)\,du\,dv}.$$

Solution with white noise

If $\{N(t)\}$ is (formal) white noise with constant spectral density N_0, the solution is simple.

Theorem 8.5. *The optimal matched filter for detection of a time-limited signal $s(t), 0 \le t \le T$, in pure white noise with spectral density N_0, has impulse response function*

$$h(u) = c\,s(T-u), \tag{8.24}$$

where $c \ne 0$ is an arbitrary constant. The optimal SNR is

$$SNR = \frac{1}{N_0}\int s(T-u)^2\,du.$$

Proof. We compute the variance of the filtered noise, using the delta function

$r_N(\tau) = N_0 \delta_0(\tau)$ as the noise covariance function, and formula (6.6):

$$E[N_{out}(T)^2] = \int_0^T \int_0^T h(u)h(v)N_0\delta_0(u-v)\,du\,dv$$

$$= N_0 \int_0^T h(u) \left\{ \int h(v)\delta_0(u-v)\,dv \right\} du = N_0 \int_0^T h(u)^2\,du,$$

and according to the Schwarz inequality,[3] the bound is

$$\text{SNR} = \frac{\{\int s(T-u)h(u)\,du\}^2}{N_0 \int h(u)^2\,du} \leq \frac{\int s(T-u)^2\,du \cdot \int h(u)^2\,du}{N_0 \int h(u)^2\,du}$$

$$= \frac{1}{N_0} \int s(T-u)^2\,du, \tag{8.25}$$

with equality if and only if $h(u) = cs(T-u)$ for some constant c. Since the right hand side in (8.25) does not depend on $h(u)$, the optimal filter has the impulse response (8.24). ∎

Solution with colored noise

To find the optimal matched filter for use in colored noise, one has to solve a certain integral equation.

Theorem 8.6. *The impulse response $h(u)$ of the optimal matched filter for detection of a time-limited signal $s(t), 0 \leq t \leq T$, in colored noise with covariance function $r_N(\tau)$, is a solution to the integral equation*

$$s(T-u) = c \int h(v)r_N(u-v)\,dv, \tag{8.26}$$

for some arbitrary constant $c \neq 0$, and the optimal SNR is

$$\text{SNR} = c^2 \int\int h(u)h(v)r_N(u-v)\,du\,dv.$$

Proof. We need a generalization of the Schwarz inequality that says, that for any covariance function $r(\tau)$ and functions $h(u), g(u)$, the inequality

$$\left(\int\int h(u)g(v)r(u-v)\,du\,dv \right)^2 \leq \tag{8.27}$$

$$\int\int h(u)h(v)r(u-v)\,du\,dv \cdot \int\int g(u)g(v)r(u-v)\,du\,dv$$

---

[3] $(\int f(x)g(x)\,dx)^2 \leq \int f^2(x)\,dx \int g^2(x)\,dx.$

holds, with equality if and only if $h(u) = c\,g(u)$, for some constant c. We prove this inequality at the end of the section.

Now, let g be a solution to the integral equation

$$s(T - u) = \int g(v) r_N(u - v)\, dv. \tag{8.28}$$

The signal-to-noise ratio can then be calculated as

$$\mathrm{SNR} = \frac{\{\int s(T-u)h(u)\,du\}^2}{\int\int h(u)h(v)r_N(u-v)\,du\,dv} = \frac{\{\int\int g(v)r_N(u-v)h(u)\,du\,dv\}^2}{\int\int h(u)h(v)r_N(u-v)\,du\,dv},$$

and according to (8.27) it is bounded as

$$\mathrm{SNR} \le \frac{\int\int g(u)g(v)r_N(u-v)\,du\,dv \cdot \int\int h(u)h(v)r_N(u-v)\,du\,dv}{\int\int h(u)h(v)r_N(u-v)\,du\,dv}$$

$$= \int\int g(u)g(v)r_N(u-v)\,du\,dv,$$

with equality if and only if $h(u) = c\,g(u)$. Thus every filter that maximizes SNR is proportional to $g(u)$, and solves the integral equation (8.26). ∎

Note that the solution reduces to the white noise solution if $r_N(\tau) = N_0\delta_0(\tau)$, since then

$$s(T - u) = c \int h(v) N_0 \delta_0(u - v)\, dv = c' h(u),$$

with $c' = c N_0$.

If the noise is almost white with approximately constant spectral density N_0, and $\int r_N(u)\,du = N_0$ and $r_N(u) \approx 0$ for $|u| > \varepsilon$, some small value, then

$$\int h(v) r_N(u - v)\, du \approx N_0 h(u).$$

Therefore, (8.24) gives an approximate solution to (8.26), and the filter (8.24) is almost optimal. Thus, our formal calculations give a reasonable result in the case of white noise.

Example 8.10. We give an example of the difference between a naive detector and the optimal detector with white or colored noise, for a simple signal in discrete time. The signal and covariance function of the noise are assumed to be, respectively,

$$s(u) = \begin{cases} 1 & \text{for } u = 0, \\ 2 & \text{for } u = 1, \\ 3 & \text{for } u = 2, \\ 2 & \text{for } u = 3, \\ 1 & \text{for } u = 4, \end{cases} \qquad r_N(\tau) = \begin{cases} 1 & \text{for } \tau = 0, \\ 0 & \text{for } \tau = 1, \\ -1/2 & \text{for } \tau = 2, \\ 0 & \text{for } \tau = 3, \\ 1/4 & \text{for } \tau = 4. \end{cases}$$

One naive detector is the sum $Y_0(4) = \sum_0^4 X(t)$, neglecting covariances and the special form of the signal. The corresponding impulse response function is $h_{naive}(u) = 1, u = 0, \ldots, 4$. The detector that would be optimal if the noise had been white has impulse response $h_{white}(u) = s(4-u)$, while the optimal filter for the assumed noise covariance is the solution to the linear equation system, $s(T-u) = \sum_0^4 h(v) r_N(u-v)$, for $u = 0, \ldots, 4$, which is the discrete time analogue of (8.26),

$$
\begin{pmatrix} s(0) \\ s(1) \\ s(2) \\ s(3) \\ s(4) \end{pmatrix} = \begin{pmatrix} 1 \\ 2 \\ 3 \\ 2 \\ 1 \end{pmatrix} = \begin{pmatrix} 1/4 & 0 & -1/2 & 0 & 1 \\ 0 & -1/2 & 0 & 1 & 0 \\ -1/2 & 0 & 1 & 0 & -1/2 \\ 0 & 1 & 0 & -1/2 & 0 \\ 1 & 0 & -1/2 & 0 & 1/4 \end{pmatrix} \begin{pmatrix} h(0) \\ h(1) \\ h(2) \\ h(3) \\ h(4) \end{pmatrix}.
$$

The solution is $h = \frac{1}{3}(10, 12, 19, 12, 10)'$, and hence the three impulse functions have the coefficients,

$$
\begin{aligned}
h_{naive} &= (1, 1, 1, 1, 1)', \\
h_{white} &= (1, 2, 3, 2, 1)', \\
h_{opt} &= (10, 12, 19, 12, 10)',
\end{aligned}
$$

neglecting any irrelevant constant. The three filter outputs are

$$
\begin{aligned}
Y_{naive}(T) &= X_0 + X_1 + X_2 + X_3 + X_4, \\
Y_{white}(T) &= X_0 + 2X_1 + 3X_2 + 2X_3 + X_4, \\
Y_{opt}(T) &= 10X_0 + 12X_1 + 19X_2 + 12X_3 + 10X_4,
\end{aligned}
$$

and the SNR is equal to 32.4, 38, and 42, with error rate, 0.22%, 0.1%, and 0.05%, respectively.   ▲

Remark 8.3. *In our search for an optimal filter we decided on a decision time T and found a solution such that, for Gaussian disturbances, all information that is needed to discriminate, in an optimal way, between the two cases "signal sent" and "no signal sent," is collected in the single number $Y(T) = \int X(T-u) h(u) \, du$. This means that we have designed a statistical test between two mean value functions, $m(t) = s(t), 0 \le t \le T$, or $m(t) = 0$.*

One may ask if one could have done better by another choice of decision time $T' < T$ with another test quantity $Y'(T')$, or, even better, with a linear combination of $Y(T)$ and $Y'(T')$. The answer is no! The reason is that a solution for T' is also a solution for T but with $h(u) = 0$ for $0 < u < T - T'$.

Proof of inequality (8.27)

We use the same technique as in the proof of the simple Schwarz' inequality. If a stochastic process with covariance function $r(\tau)$ passes through a linear filter with impulse response function $h(t) - cg(t)$, the variance of the output is always non-negative, i.e., by Formula (6.6) for the variance of $\int (h(u) - cg(u))X(t - u)\,du$,

$$
0 \le \iint (h(u) - cg(u))(h(v) - cg(v))r(u - v)\,du\,dv
$$

$$
= \iint h(u)h(v)r(u - v)\,du\,dv
$$

$$
- 2c \iint h(u)g(v)r(u - v)\,du\,dv + c^2 \iint g(u)g(v)r(u - v)\,du\,dv.
$$

Dividing by $2c$ and moving the middle term to the left hand side we get

$$
\iint h(u)g(v)r(u - v)\,du\,dv
$$

$$
\le \frac{1}{2}\left(\frac{1}{c}\iint h(u)h(v)r(u - v)\,du\,dv + c\iint g(u)g(v)r(u - v)\,du\,dv\right),
$$

and (8.27) follows if we take

$$
c = \frac{\sqrt{\iint h(u)h(v)r(u - v)\,du\,dv}}{\sqrt{\iint g(u)g(v)r(u - v)\,du\,dv}}.
$$

The equality for $h(u) = cg(u)$ follows directly by insertion into (8.27).

8.5 Wiener filter

The Wiener filter is one of the most important devices in stochastic process applications. It was derived by Norbert Wiener in 1940, and independently by A.N. Kolmogorov in 1941, as a general method to clean a desired random signal from unwanted random noise.

In many applications one wants to separate an interesting, but stochastic, signal from uninteresting noise. The signal may be a piece of music, part of a speech, or a segment of a picture. It may even be the future part of an observed random process; then one talks about an "optimal predictor."

It is typical that the exact shape of the signal is not known, not even determined; only its correlation or spectral properties are supposed to be known. A short piece of music from a vinyl recording, including surface noise, is a good example. Note the difference to the situation in the previous section, where the signal was deterministic, and its shape was known by the receiver; only its presence or absence remained to be determined.

8.5.1 Reconstruction of a stochastic process

The Wiener filter is designed to present an estimate of or to reconstruct an interesting stochastic signal from noisy data. In the most simple case we want to clean a stochastic information signal $\{S(t), t \in \mathbb{R}\}$ from a disturbing, like-wise stochastic, noise $\{N(t), t \in \mathbb{R}\}$, with different spectral content. This is the "vinyl cleaning example." In a more general situation, there is some other stochastic signal $\{D(t), t \in \mathbb{R}\}$, called "the desired signal," which is related to, but perhaps not exactly equal to, $\{S(t), t \in \mathbb{R}\}$, and we want to estimate $D(t)$ from a noisy observation of $\{S(t), t \in \mathbb{R}\}$. Continuing the vinyl example, one can think of $\{D(t), t \in \mathbb{R}\}$ as the cello tone and $\{S(t), t \in \mathbb{R}\}$ as the full orchestra, and we want to filter out just the cello in a noisy recording. By "optimal" we mean that the filter minimizes the variance of the difference between the desired signal $D(t)$ and the reconstruction, as represented by the filter output $Y(t)$. Figure 8.8 illustrates the principle.

The optimality criterion, "minimal squared error between desired signal and the reconstructed signal," is quite different from the criterion used for the matched filter. There, it was obvious that the goal was "small error probabili-ties," and as we saw, that meant maximizing a specific "signal-to-noise" ratio. To evaluate a cleaning filter, one has to adopt a more general type of statisti-cal criterion, which works on the average, taking care also of the randomness in the interesting signal. The common way to define such a signal-to-noise criterion is the following more general version of Definition 8.3.

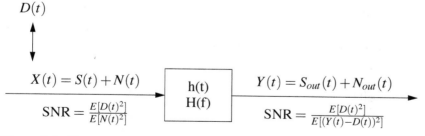

Figure 8.8 *Cleaning of stochastic signal $S(t)$ disturbed by random noise $N(t)$.*

> **Definition 8.4.** *In a signal processing system, the signal-to-noise ratio is defined as*
>
> $$\text{SNR} = \frac{average\ effect\ of\ desired\ signal}{average\ error\ effect}.$$

The definition is based on quantities that are easy to measure, and to handle mathematically. In an audio application, it is also related to how well the human ear perceives noise disturbances: this is usually assumed to be proportional to the logarithm of the SNR (dB). (Of course this can only be approximatively correct, since the sensitivity of the ear is frequency dependent, and since some types of noise are more disturbing than others.)

8.5.2   Optimal cleaning filter; frequency formulation

We start with the simple problem to remove noise from a stochastic signal, and present the solution in the frequency domain. The more general problem, to reconstruct any interesting signal from noisy observations of some other signal, will be treated in the next section, Section 8.5.3.

Suppose the interesting signal is a stationary stochastic process $\{S(t)\}$ with mean 0, covariance function $r_S(\tau)$, and spectral density $R_S(f)$. Also suppose the signal is disturbed by a stationary stochastic noise process $\{N(t)\}$, independent of $\{S(t)\}$, with mean 0, covariance function $r_N(\tau)$, and spectral density $R_N(f)$. Our task is to find an impulse response $h(t)$ and frequency function $H(f)$ that suppresses the noise as much as possible. By that we mean that the output $Y(t)$ from the filter shall be as close to $S(t)$ as possible (in the mean square sense), when the input is $X(t) = S(t) + N(t)$.

We denote the filtered signal by $S_{out}(t)$ and the filtered noise by $N_{out}(t)$, so that the output from the filter is $Y(t) = S_{out}(t) + N_{out}(t)$; see Figure 8.8. The error in the reconstruction is $Y(t) - S(t) = S_{out}(t) + N_{out}(t) - S(t)$, with mean square error

$$\mathsf{E}[(Y(t) - S(t))^2] = \mathsf{E}[(S_{out}(t) + N_{out}(t) - S(t))^2]$$
$$= \mathsf{E}[(S_{out}(t) - S(t))^2] + \mathsf{E}[N_{out}(t)^2], \tag{8.29}$$

since we have assumed that signal and noise are independent processes. We seek $H(f)$ that minimizes the reconstruction error (8.29), which obviously is equivalent to maximizing the SNR, defined by Definition 8.4, i.e.,

$$\text{SNR} = \frac{\text{E}[S(t)^2]}{\text{E}[(Y(t) - S(t))^2]}, \tag{8.30}$$

since the numerator $\text{E}[S(t)^2]$ is independent of the filter characteristic $H(f)$.

Theorem 8.7 ("Wiener filter"). *The optimal cleaning filter that minimizes the mean square reconstruction error and maximizes the SNR (8.30) has frequency function*

$$H(f) = \frac{R_S(f)}{R_S(f) + R_N(f)}, \tag{8.31}$$

where $R_S(f)$ and $R_N(f)$ are the signal and noise spectral densities, respectively.

The optimal signal-to-noise ratio is

$$\text{SNR}_{max} = \frac{\int R_S(f)\,df}{\int \frac{R_S(f)R_N(f)}{R_S(f)+R_N(f)}\,df}. \tag{8.32}$$

Proof. We shall find the filter that minimizes the sum of the two terms in (8.29). For the first term, we observe that since

$$S_{out}(t) - S(t) = \int S(t-u)h(u)\,du - S(t) = \int S(t-u)(h(u) - \delta_0(u))\,du$$

is the output from a filter with input signal $\{S(t)\}$ and frequency function $\int_{-\infty}^{\infty} e^{-i2\pi ft}(h(t) - \delta_0(t))\,dt = H(f) - 1$, Theorem 6.2, in Section 6.2, gives

$$\text{E}[(S_{out}(t) - S(t))^2] = \int_{-\infty}^{\infty} |H(f) - 1|^2 R_S(f)\,df.$$

Similarly, with input $\{N(t)\}$ and output $\{N_{out}(t)\}$,

$$\text{E}[N_{out}(t)^2] = \int_{-\infty}^{\infty} |H(f)|^2 R_N(f)\,df,$$

and adding the two terms gives

$$\text{E}[(Y(t) - S(t))^2] = \int_{-\infty}^{\infty} \left\{ |H(f) - 1|^2 R_S(f) + |H(f)|^2 R_N(f) \right\} df. \tag{8.33}$$

LINEAR FILTERS – APPLICATIONS

Since both R_S and R_N are real and non-negative, the integrand in (8.33) is always non-negative, and we can minimize the integral by minimizing the integrand for each f, separately. Suppressing the argument f, we can expand in real ($\Re H$), and imaginary ($\Im H$), parts,

$$
\begin{aligned}
|H-1|^2 R_S + |H|^2 R_N &= (H-1)(\overline{H}-1)R_S + H\overline{H}R_N \\
&= H\overline{H}(R_S+R_N) - (H+\overline{H})R_S + R_S \\
&= (\Re H)^2(R_S+R_N) + (\Im H)^2(R_S+R_N) - 2(\Re H)R_S + R_S.
\end{aligned}
$$

To minimize this, take $\Im H = 0$, and choose $H = \Re H$, to minimize

$$
H^2(R_S+R_N) - 2HR_S + R_S,
$$

for real H. If we assume that the spectral densities are regular functions, we find by differentiation that the minimum is attained for $H = R_S/(R_S+R_N)$, and the minimum value of the integral is

$$
\int \frac{R_S(f)R_N(f)}{R_S(f)+R_N(f)}\, df,
$$

which is the statement of the theorem.   ∎

Figure 8.9 illustrates the optimal filter for two different signal and noise spectra. For frequencies where the signal dominates the noise, the damping from $H(f) \approx 1$ is small, while in the opposite case $H(f) \approx 0$, and the damping is substantial. For f such that $R_S(f) + R_N(f) = 0$, the filter can be chosen arbitrarily. A drawback with the optimal filter is that it is not causal, and for real-time applications one has to find an approximation.

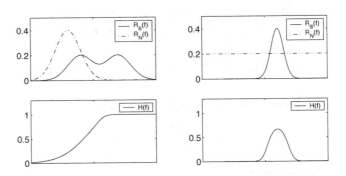

Figure 8.9 *Noise spectrum $R_N(f)$, signal spectrum $R_S(f)$, and frequency function $H(f)$ for the Wiener filter in two cases of signal/noise spectrum.*

8.5.3   General solution; discrete time formulation

We now turn to the general problem of reconstructing a "desired random signal" from observations of a related stochastic process. In order to throw further light on the problem we will seek the solution in the time domain instead of in the frequency domain, and to obtain an explicit solution, we formulate it in discrete time.

Let $\{D_n, n \in \mathbb{Z}\}$ be the desired, unobserved signal that we want to reconstruct, based on observations of some other process $\{X_n, n \in \mathbb{Z}\}$ with covariance function $r_X(k) = C[X_n, X_{n+k}]$ and cross-covariance function $r_{XD}(k) = C[X_n, D_{n+k}]$. We assume both sequences have expectation zero.

We formulate the following theorem, extending Theorem 8.7.

Theorem 8.8. *(a) The impulse function $\{h_k\}$ of the optimal reconstruction filter $\widehat{D}_n = \sum_j h_j X_{n-j}$ satisfies the infinite equation system*

$$\sum_j h_j r_X(j - k) = r_{XD}(k), \quad k \in \mathbb{Z}. \tag{8.34}$$

(b) The frequency function of the optimal reconstruction filter is

$$H(f) = \frac{R_{XD}(f)}{R_X(f)}, \tag{8.35}$$

where $R_{XD}(f)$ and $R_X(f)$ are the cross-spectral and (auto) spectral densities, respectively.

(c) In the special case $D_n = S_n$ with observations disturbed by noise, uncorrelated with $\{S_n\}$, i.e., $X_n = S_n + N_n$, then $r_{XD}(n) = r_S(n)$, $r_X(k) = r_S(k) + r_N(k)$, and the optimal filter has frequency function

$$H(f) = \frac{R_S(f)}{R_S(f) + R_N(f)}, \tag{8.36}$$

in agreement with Theorem 8.7.

Proof. We seek filter coefficients h_k, with $\sum_k h_k^2 < \infty$, such that the error signal,

$$E_n = D_n - Y_n = D_n - \sum_{k=-\infty}^{\infty} h_k X_{n-k},$$

has minimal variance, i.e., $\{h_k\} = \arg\min_{h_k} E[E_n^2]$, since the error has mean zero. Here, since neither D_n nor the X_{n-j} depend on h_k,

$$\frac{\partial}{\partial h_k} E[E_n^2] = E\left[2E_n \frac{\partial}{\partial h_k} E_n\right]$$

$$= 2E\left[E_n\left\{\frac{\partial}{\partial h_k} D_n - \frac{\partial}{\partial h_k} h_k X_{n-k} - \sum_{j \neq k}\left(\frac{\partial}{\partial h_k} h_j\right) X_{n-j}\right\}\right] = -2E[E_n X_{n-k}].$$

To find the minimum, we set the derivatives equal to zero, and obtain the orthogonality property for the error terms,

$$E[E_n X_{n-k}] = r_{XD}(k) - \sum_j h_j r_X(j-k) = 0$$

for all n and k. Note that this agrees with the general orthogonality principle (7.7). ∎

8.6 Kalman filter

"The" Kalman filter is a generic name for a class of filters which are widely used in control theory, signal processing and econometrics. It is named after Rudolf E. Kalman,[4] one of its several originators, [28]. The basic concept underlying the Kalman filter is a *state space model* of the processes of interest, see (8.37) and (8.38) below, and the key result is that this representation makes it possible to give simple *recursive* formulas for prediction, estimation, and smoothing. The state space representation is quite general, and in particular can be seen to include the ARMA-processes treated in Chapter 7, and it also has important extensions to non-stationary processes.

For ease of presentation we use the simplest one-dimensional version of the state space representation and the Kalman filter. However, the formulas we obtain are in fact generally valid, and the derivations for the more important and interesting multi-dimensional versions are the same as the derivation below; one just has to replace the sequences $\{X_t\}$ and $\{Y_t\}$ by sequences of vectors, and the constants A and C by matrices, and division by matrix inverses, or, if the matrices are not invertible, by generalized matrix inverses.

8.6.1 The process and measurement models

The state space representation of a system uses a recursive model, the *state space equation*,

$$X_{t+1} = AX_t + e_{t+1} \tag{8.37}$$

---

[4]Rudolf E. Kalman, Hungarian born mathematician and electrical engineer, 1930 –.

for the *states* $\{X_n, n \in \mathbb{Z}\}$ of the system, and an *observation equation,*

$$Y_t = CX_t + f_t \tag{8.38}$$

for *the observed values* $\{Y_n, n \in \mathbb{Z}\}$. Typically the states X_t are the basic interesting physical or macroeconomic properties of a system, e.g. the speed of a vehicle at time t, while the Y_t are the values which one has measurements of, say, the location of the vehicle at time t. Further, we assume that $\{e_n, n \in \mathbb{Z}\}$ is a sequence of independent normal variables, $e_t \sim N(0, \sigma_e^2)$ and $\{f_n, n \in \mathbb{Z}\}$ are independent normal errors, $f_t \sim N(0, \sigma_f^2)$, and that A (with $|A| < 1$), and C are constant scalars. In the terminology of Chapter 7, $\{X_t\}$ is an AR(1)-process.

Denote the vector of observed values obtained up till time t by

$$\mathcal{Y}_t = (Y_1, \ldots, Y_t),$$

and let

$$\widehat{X}_{t+k|t} = E[X_{t+k} \mid \mathcal{Y}_t]$$

for $k = 0, 1, \ldots$. Here $\widehat{X}_{t|t}$ is an estimator (or "reconstructor") of the state of the system at time t, given all the information available at time t, and $\widehat{X}_{t+k|t}$, for $k = 1, 2, \ldots$, are predictors of future states of the system, given the information available at time t. Below we will derive simple recursive formulas for the estimator and predictors of the systems. These formulas allow for very fast computation, and in particular are suited for real-time applications.

The derivation will also give formulas for the predictors

$$\widehat{Y}_{t+k|t} = E[Y_{t+k} \mid \mathcal{Y}_t], \; k = 1, 2, \ldots,$$

of future observations, given the information available at time t.

8.6.2 *Updating a conditional normal distribution*

The Kalman filter is based on a formula for updating the conditional mean and covariances in a multivariate normal distribution when new information becomes available. In the formula, $\mathcal{Y}$ is a vector of "old" random variables, Y is the random variable that contains the "new" information, and we assume that X, Y, and $\mathcal{Y}$ have a joint multivariate normal distribution. The updating equation then computes the new expected value and variance of X from the old conditional expected value $E[X \mid \mathcal{Y}]$, the old conditional variance $V[X \mid \mathcal{Y}]$, and the old conditional covariance $C[X, Y \mid \mathcal{Y}]$ as follows,

$$\begin{aligned} E[X \mid Y, \mathcal{Y}] &= E[X \mid \mathcal{Y}] + C[X, Y \mid \mathcal{Y}] V[Y \mid \mathcal{Y}]^{-1} (Y - E[Y \mid \mathcal{Y}]), \\ V[X \mid Y, \mathcal{Y}] &= V[X \mid \mathcal{Y}] - C[X, Y \mid \mathcal{Y}] V[Y \mid \mathcal{Y}]^{-1} C[X, Y \mid \mathcal{Y}]'. \end{aligned} \tag{8.39}$$

These formulas follow from Theorem A.3 by noting that the joint distribution of X and Y conditional on $\mathscr{Y}$ is a bivariate normal distribution where the expected value of X is $\mathsf{E}[X \mid \mathscr{Y}]$, the variance of X is $\mathsf{V}[X \mid \mathscr{Y}]$, and the covariance between X and Y is $\mathsf{C}[X,Y \mid \mathscr{Y}]$.

8.6.3   The Kalman filter

We now apply (8.39) with $X = X_t$, $Y = Y_t$, $\mathscr{Y} = \mathscr{Y}_{t-1}$ to obtain recursive equations for computing estimators and predictors for the state space model (8.37) and (8.38). For this we use the further notation

$$V_{XX}[t+k \mid t] = \mathsf{V}[X_{t+k} \mid \mathscr{Y}_t], \qquad V_{YY}[t+k \mid t] = \mathsf{V}[Y_{t+k} \mid \mathscr{Y}_t],$$
$$C_{XY}[t+k \mid t] = \mathsf{C}[X_{t+k},Y_{t+k} \mid \mathscr{Y}_t],$$

and

$$\widetilde{X}_{t+k|t} = X_{t+k} - \widehat{X}_{t+k|t}, \qquad \widetilde{Y}_{t+k|t} = Y_{t+k} - \widehat{Y}_{t+k|t},$$

and define the *Kalman gain* as

$$K(t) = C_{XY}(t|t-1)V_{YY}^{-1}(t|t-1).$$

With this notation (8.39) gives that the state estimator $\widehat{X}_{t|t}$ can be computed recursively using the formulas

$$\begin{aligned}
\widehat{X}_{t|t} &= \widehat{X}_{t|t-1} + K(t)\left(Y_t - \widehat{Y}_{t|t-1}\right), \\
V_{XX}(t|t) &= V_{XX}(t|t-1) - K(t)V_{YY}(t|t-1)K(t)'.
\end{aligned} \tag{8.40}$$

Further, since the variables $e_{t+1}, e_{t+2}, \dots$ in Equation (8.37) are independent of the values $\mathscr{Y}_t$ which have been observed up to time t, it is clear that

$$\widehat{X}_{t+1|t} = A\widehat{X}_{t|t} \quad \text{and} \quad \widehat{X}_{t+k|t} = A^k\widehat{X}_{t|t}, \quad \text{for} \quad k = 2,3,\dots. \tag{8.41}$$

Similarly, since $f_{t+1}, f_{t+2}, \dots$ are independent of the information available up to time t, it also follows that

$$\widehat{Y}_{t+1|t} = C\widehat{X}_{t+1|t}. \tag{8.42}$$

Equation (8.39) additionally provides formulas for recursive computation of the Kalman gain and of the variances in (8.40),

$$\begin{aligned}
V_{XX}(t+1|t) &= AV_{XX}(t|t)A' + \sigma_e^2, \\
V_{YY}(t|t-1) &= CV_{XX}(t|t-1)C' + \sigma_f^2, \\
C_{XY}(t|t-1) &= V_{XX}(t|t-1)C',
\end{aligned} \tag{8.43}$$

where we have used that since the innovations e_{t+1} and measurement errors f_{t+1} are independent of $\mathscr{Y}_t$ and hence of the values predicted at time t it holds that

$$\widetilde{X}_{t+1|t} = X_{t+1} - \widehat{X}_{t+1|t} = A\widetilde{X}_{t|t} + e_{t+1},$$
$$\widetilde{Y}_{t|t-1} = Y_t - \widehat{Y}_{t|t-1} = C\widetilde{X}_{t|t-1} + f_t.$$

To start the recursions one uses that at time $t = 0$ the conditional moments are equal to the unconditional ones.

Theorem 8.9 ("Kalman filter"). *The Kalman filter for the state space model* (8.37) *and* (8.38) *is the collection of recursive equations* (8.41)–(8.43). *It minimizes the variance of the estimation and prediction errors.*

To prove the last part of the theorem, first note that for multivariate normal distributions the conditional means have smallest error variance, and then that the error variances only depend on the second moments of the processes, so that the Kalman filter gives the smallest error variances for any second order process. In fact, Kalman filters often are used also for non-normal processes.

8.7 An example from structural dynamics

We end this chapter with a simple example from structural stochastic dynamics, one of the areas where stochastic processes introduced a new way of thinking in the aerospace, automotive, and marine industry, in the middle of the last century. A wide range of phenomena were described in stochastic terms, in contrast to the deterministic formulations, which had been dominating, based on the Newtonian mechanics. Stochastic process models were formulated for ocean waves, wind turbulence, rocket engine noise, earthquake vibrations, not to speak about the development of stochastic control theory, which made space technology possible.

An offshore oil platform can be thought of as a linear or non-linear filter, which takes waves from the random sea as input and transforms them into varying stress levels on critical construction details in the platform, leading to random fatigue. A corrugated, uneven rail track can cause random vibrations of a train engine, depending on the space frequency of the corrugation, etc.

We shall now describe a simplified mechanical system, as an introduction to the type of problems one can handle.

A car or truck is a flexible body, with wheels attached to the body by springs and dampers. It travels on an uneven road, which can be thought of as a random surface, which is stationary, at least locally. The covariance function and spectral density of the surface deviation from average height has to be chosen in accordance with the realistic hills, pits and holes, and fine gravel or asphalt structure of the road. For examples of road models used by truck manufacturers, see [8].

The suspension system acts as a filter, with the time signal from the road as input, and if the excitation is moderate, the filter can be assumed to be linear. For a statistically homogeneous road, the movements of the car will then be a stationary stochastic process. We will now with a simple example illustrate the use of the statistical model.

8.7.1   A one-wheeled car

Figure 8.10 symbolizes a car with one, very small, weightless wheel, attached by a spring and damper system. The spring and damping constants are denoted by k and c, respectively. The total mass of the car is m, and the unloaded length of the spring is y_0. The height of the road surface above the average level is denoted by $\widetilde{X}(s)$, where s is the distance from the starting point.

On its route along the road, the car experiences random vertical vibrations, and the suspension system tries to keep these as small as possible, and at the same time keep the wheel in safe contact with the road. We shall derive the basic statistical properties of these comfort and safety variables, under the assumption that the car drives with constant speed, v.

Let $X(t)$ be the road surface height at *time* t, after start, i.e.,

$$X(t) = \widetilde{X}(vt), \tag{8.44}$$

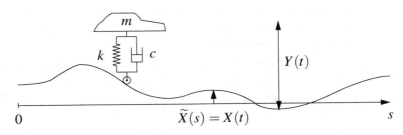

Figure 8.10 *Road and car at location $s = vt$.*

and let $Y(t)$ be the extension of the spring, equal to the height of the car above the surface, under the somewhat unrealistic assumption that the wheel never leaves the ground. Then the following differential equation governs the relation between $X(t)$ and $Y(t)$,

$$m(Y''(t) + X''(t)) + cY'(t) + k(Y(t) - y_0) = -gm. \qquad (8.45)$$

Here, $g = 9.81$ ms^{-2} is the earth acceleration constant, and gm the gravity force on the car. The vertical height of the car is $X(t) + Y(t)$, the vertical acceleration $X''(t) + Y''(t)$, the expansion rate of the damper is $Y'(t)$, and the spring compression $Y(t) - y_0$, when y_0 is its unloaded length. (Note that (8.45) is only valid if the wheel is in constant contact with the road surface.) The equilibrium spring length, y_e, is given by $k(y_e - y_0) = -gm$, so equation (8.45) can be written

$$Y''(t) + \frac{c}{m}Y'(t) + \frac{k}{m}(Y(t) - y_e) = -X''(t). \qquad (8.46)$$

This equation is of the same type as the equation for the resonance circuit in Example 8.3, Equation (8.8). The only difference is that $-X''(t)$ appears in the right hand side instead of $X'(t)$.

We have described the car as a linear stochastic system with $X(t)$ as input signal and $Y(t) - y_e$ as output. The linear filter that takes $\{X(t)\}$ to $\{Y(t) - y_e\}$ has the frequency function

$$H_{XY}(f) = \frac{(2\pi f)^2}{-(2\pi f)^2 + i2\pi fc/m + k/m}. \qquad (8.47)$$

The vertical acceleration $X''(t) + Y''(t)$ is important for the comfort of driver and passengers, and in fact, attempts have been made to formulate a frequency function also for the filter from acceleration to discomfort. The frequency function from $X(t)$ to $X''(t) + Y''(t)$ is simply

$$\begin{aligned} H_{X,X''+Y''}(f) &= H_{X,X''}(f) + H_{X,Y''}(f) = -(2\pi f)^2 - (2\pi f)^2 H_{XY}(f) \\ &= -(2\pi f)^2 (1 + H_{XY}(f)). \end{aligned} \qquad (8.48)$$

8.7.2 A stochastic road model

To analyze the statistical properties of the car movements one needs a stochastic process model for the road, which depends on the type of road the car is expected to use. The simplest model used for roads, [1], has a spectral density

$$R_{\tilde{X}}(n) = R_0 n^{-2} \quad \text{for } n_{\min} \leq |n| \leq n_{\max}. \qquad (8.49)$$

Here, n is a space frequency, called *wave number*, with unit $[length\ unit]^{-1}$, for example the number of periods per meter. Common values for n_{min} and n_{max} are 0.1 m^{-1} and 100 m^{-1}, respectively, i.e., no fluctuations longer than 10 meters or shorter than 1 cm are counted. The constant R_0 in (8.49) determines the standard deviation and variance for the road profile,

$$V[\widetilde{X}(s)] = \int_{-\infty}^{\infty} R_{\widetilde{X}}(n)\,dn = 2R_0 \int_{n_{min}}^{n_{max}} n^{-2}\,dn = 2R_0 \left(n_{min}^{-1} - n_{max}^{-1}\right) = 19.8\,R_0.$$

For a reasonably bad country gravel road $R_0 \approx 5 \cdot 10^{-4}$ m^2, leading to a standard deviation of about 0.1 m.

For the dynamic road profile $X(t) = \widetilde{X}(vt)$, encountered while driving at constant speed v, the spectral density is

$$R_X(f) = v^{-1} R_{\widetilde{X}}(f/v).$$

Spectral densities for different important processes can now be found from (8.47) and (8.48). The spring length $\{Y(t)\}$ has spectral density

$$R_Y(f) = |H_{XY}(f)|^2 R_X(f) = |H_{XY}(f)|^2 \cdot v^{-1} R_{\widetilde{X}}(f/v)$$

$$= \frac{R_0 v (2\pi)^2 (2\pi f)^2}{(k/m - (2\pi f)^2)^2 + (2\pi f c/m)^2},\tag{8.50}$$

for $vn_{min} \leq f \leq vn_{max}$, that of the damper compression velocity $Y'(t)$ is

$$R_{Y'}(f) = (2\pi f)^2 R_Y(f),\tag{8.51}$$

and that of the vertical acceleration is

$$R_{X''+Y''}(f) = (2\pi f)^4 |1 + H_{XY}(f)|^2 R_X(f).$$

8.7.3 Optimization

Structural optimization often takes into account the stochastic properties of the construction, for example the expected fatigue life, and the risk (= probability) that some stress factor in the construction will exceed a specified safety limit. These properties must be evaluated, taking the expected environmental and operational conditions into account. As an example, we consider the spring and damper system and the coefficients k and c in the one-wheeled car. Then, also the *optimization criteria* have to be chosen with care.

One alternative is to reduce the vertical acceleration of the car, i.e., minimizing $V[X''(t) + Y''(t)]$, in order to optimize driving comfort. Another alternative is to maximize road holding, and minimize the variability of the normal force between wheel and road. A third alternative would be to minimize the variance $V[Y(t)]$ of the spring length. These different criteria are in conflict (small spring length variation will lead to large variations in normal force and vertical acceleration), and one has to seek a compromise.

8.8   Monte Carlo simulation in continuous time

This section deals with simulation of Gaussian processes as solutions of linear differential equations driven by white noise, like the Ornstein-Uhlenbeck process. Two methods are then available, approximate solution via the Euler scheme, $X(t+\Delta t) \approx X(t) + \Delta t X'(t)$, with generalizations to higher order, and "exact" simulation of increments.

Example 8.11 ("Ornstein-Uhlenbeck process"). The Ornstein-Uhlenbeck process,

$$X'(t) + \alpha X(t) = \sigma_0 W'(t),$$

can be simulated in its integrated form, dividing by Δ_t,

$$\frac{X(t) - X(t - \Delta_t)}{\Delta_t} = \frac{\alpha}{\Delta_t} \int_{t-\Delta_t}^{t} X(u)du + \sigma_0 \frac{W(t) - W(t - \Delta_t)}{\Delta_t}$$

$$\approx \alpha X(t) + \frac{\sigma_0}{\sqrt{\Delta_t}} N_t,$$

where N_t is a standardized normal variable, $N_t \in N(0,1)$. The Euler scheme then gives

$$X^{(a)}(t+\Delta t) = \alpha X^{(a)}(t) + \sigma_0 \sqrt{\Delta_t} N_t.$$

The exact solution is given by (8.14) and (8.15),

$$X^{(e)}(t+\Delta_t) = X^{(e)}(t)e^{-\alpha \Delta_t} + \sigma \sqrt{\frac{1 - e^{-2\alpha\Delta_t}}{2\alpha}} N_t. \qquad \blacktriangle$$

Exercises

8:1. Why is the assumption that $R_0(x)$ is zero outside the interval $|x| \le d < f_0$ important in (8.21)?

8:2. A stationary sequence $\{S(t), t = 0, \pm 1, \pm 2 \ldots\}$ has expectation zero and covariance function,

$$r_S(\tau) = \begin{cases} 2, & \tau = 0 \\ -1, & |\tau| = 1 \\ 0, & \text{otherwise.} \end{cases}$$

The process is disturbed by colored noise $\{N(t), t = 0, \pm 1, \pm 2 \ldots\}$ with covariance function,

$$r_N(\tau) = \begin{cases} 2, & \tau = 0 \\ 1, & |\tau| = 1 \\ 0, & \text{otherwise.} \end{cases}$$

The process and the disturbance are independent. Determine the frequency function $H(f)$, $-1/2 < f \le 1/2$, and the impulse response $h(t)$, $t = 0, \pm 1, \pm 2 \ldots$, for the optimal Wiener filter that minimizes $E[(Y(t) - S(t))^2]$ when $Y(t)$ is the output signal from the filter with $X(t) = S(t) + N(t)$ as input signal.

8:3. Suppose that we want to measure the temperature, $X(t)$, in a manufacturing process. Since the measurements are corrupted by additive noise, $N(t)$, independent of the temperature, we reconstruct the true temperature with an optimal filter. Assuming $r_X(\tau) = e^{-2|\tau|}$ and $r_N(\tau) = e^{-20|\tau|}$, find the covariance function of the output signal, $\{Y(t)\}$.

8:4. Suppose we want to transmit a bandlimited stochastic signal, $S(t)$, on a channel disturbed by flicker noise. The spectral density of the signal is

$$R_S(f) = \begin{cases} 1, & 100 \le |f| \le 1000, \\ 0, & \text{otherwise,} \end{cases}$$

and that of the flicker noise is $R_N(f) = 100/|f|$ up to high frequencies.

(a) Design an optimal noise reducing filter. Make a plot of the frequency response and compute the SNR (signal-to-noise ratio).

(b) The following transfer function is not optimal,

$$G(f) = \begin{cases} 1, & 100 \le |f| \le 1000, \\ 0, & \text{otherwise.} \end{cases}$$

Compute the SNR of the output signal with this filter.

8:5. An AR(2)-process $\{X_t\}$ has a covariance function, estimated to $\widehat{r}_X(0) = 4$, $\widehat{r}_X(\pm 1) = 2$, and $\widehat{r}_X(\pm 2) = -0.5$. Determine from these values, the parameters for the AR-process, and find the frequency function (in real-valued form) for a Wiener filter which in the best way reconstructs X_t from the measured signal $Y_t = X_t + N_t$, when N_t is white noise with expected value 0 and variance 1. Assume that $\{X_t\}$ and $\{N_t\}$ are independent.

8:6. A deterministic signal, $s(t)$, is corrupted by additive white noise with spectral density N_0. The signal is defined by

$$s(t) = \begin{cases} Ae^{-bt}, & t \ge 0 \\ 0, & \text{otherwise.} \end{cases}$$

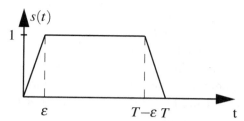

Figure 8.11 *The signal, s(t) in Exercise 8:7.*

The signal and the noise is processed by a linear optimal filter. We read the output of this filter at time $t = T > 0$. The filter has the property that the SNR is maximized at $t = T$.

(a) Determine the filter and the maximal SNR.

(b) The filter used in (a) is non-causal. Make a small adjustment in order to change this, and compute the new SNR.

(c) How large must T be, in order for the SNR to be larger than 99% of the maximum SNR?

8:7. A sender transmits binary characters to a receiver in a digital communication system. The signal, $s(t)$, see Figure 8.11, is used by the sender to represent a binary one and the absence of this signal is used to represent zero. We model the received signal, $X(t)$, by

$$X(t) = \text{sent signal} + N(t), \quad 0 \le t \le T,$$

where $\{N(t), t \in \mathbb{R}\}$ is Gaussian noise with spectral density R_0. Determine the impulse response of a matched causal filter and compute the SNR (signal-to-noise ratio) of the output signal.

8:8. A deterministic signal, s_k, in discrete time, is given by

$$s_k = \begin{cases} 1, & k = 0, 1, 2, 3, \\ 0, & \text{otherwise.} \end{cases}$$

We would like to determine whether this signal has been sent or not, when one receives the signal $s_k + N_k$ if the signal was sent and N_k if it was not. We assume that N_k is standardized white Gaussian noise.

(a) We use a filter, defined by its impulse response:

$$h(k) = \begin{cases} 1, & k = 2, 3, \\ 0, & \text{otherwise} \end{cases}$$

and choose the decision time $k = 3$. If the output of the filter is larger than 1 at $k = 3$ we consider the signal to be present. What is the probability that we erroneously believe that the signal has been sent?

(b) The filter used in (a) is not optimal. Design an optimal causal filter and specify the time of decision and the threshold level of the discriminator such that the probabilities of the two possible errors are equal. What is the probability of an error?

8:9. A weakly stationary process $\{S_t, t = 0, \pm 1, \pm 2 \ldots\}$ has mean zero and covariance function,

$$r_S(\tau) = \begin{cases} 2, & \tau = 0, \\ -1, & |\tau| = 1, \\ 0, & \text{otherwise.} \end{cases}$$

The process is disturbed by colored noise $\{N_t, t = 0, \pm 1, \pm 2 \ldots\}$ with expected value zero and covariance function

$$r_N(\tau) = \begin{cases} 2, & \tau = 0, \\ 1, & |\tau| = 1, \\ 0, & \text{otherwise.} \end{cases}$$

The process and the disturbance are independent. In communication systems it is quite usual with echo, i.e., you have a signal according to

$$X_t = S_t + 0.5 S_{t-1} + N_t,$$

where $0.5 S_{t-1}$ is the damped and delayed original process realization. Determine the coefficients a, b for the non-causal impulse response

$$h(u) = \begin{cases} b, & u = -1, \\ a, & u = 0, \\ 0, & \text{otherwise.} \end{cases}$$

for an optimal filter that minimizes $E[(Y_t - S_t)^2]$, when Y_t is the output signal and X_t is the input signal to the optimal filter.

8:10. Find explicit expressions for the Kalman filter for an AR(1)-process observed with measurement errors,

$$X_t = \phi_1 X_{t-1} + e_t,$$
$$Y_t = X_t + f_t,$$

when $\{e_t\}$ and $\{f_t\}$ are independent normal variables with mean zero and $V[e_t] = \sigma_e^2$, $V[f_t] = \sigma_f^2$.

Chapter 9

Frequency analysis and spectral estimation

9.1 Introduction

Estimation of a spectrum from a small number of data samples is a classical research field. Traditionally, spectral analysis of a sampled stationary time series is formulated as either a parametric or a non-parametric estimation problem, or as a hybrid of the two. The windowed periodogram (and lag windowed covariance estimates) represents the *non-parametric* approach. The *parametric* approach is represented by various forms of autoregressive (AR)-based methods. In fact, a combination of the two methods often gives the best results.

In 1704 Isaac Newton[1] performed an experiment, using a glass prism to resolve the sunbeams into the colors of the rainbow, a spectrum. In 1807 Fourier[2] suggested expressing a possibly discontinuous function as a sum, a spectrum, of continuous functions. This seemingly absurd idea was eventually accepted and is what we today call the Fourier expansion.

The modern history of spectral estimation begins with J.W. Tukey[3] in 1949 but a first step was made already in 1898 by Arthur Schuster.[4] He fitted Fourier series to sunspot numbers to find hidden periodicities and thereby the periodogram was invented, [40]. The method has been frequently used for spectrum estimation ever since Cooley[5] and Tukey (re)invented the Fast Fourier Transform (FFT) algorithm in 1965, [15]. The FFT is a way to calculate the Fourier transform of a discrete time signal, which exploits the structure of the Fourier transform algorithm to minimize the number of multi-

---

[1] Sir Isaac Newton, English mathematician, physicist, etc., 1642 – 1726.
[2] Jean-Baptiste Joseph Fourier, French mathematician and physicist, 1768 – 1830.
[3] John Wilder Tukey, American mathematician, 1915 – 2000.
[4] Sir Arthur Schuster, German-born British physicist, 1851 – 1934.
[5] John William Cooley, American mathematician, 1926 – .

plications and summations. This allowed the Fourier transform to become a practically useful tool, and not just a theoretical description.

The chapter starts with the periodogram as the basic tool for estimation of the spectral density of a stationary process. The expectation and variance are derived in Section 9.2 and it is shown that the relative standard deviation of the periodogram estimator is asymptotically constant as the sample size increases. Section 9.3 deals with the FFT-algorithm.

Sections 9.4 and 9.5 present different methods to reduce the variance of the spectral estimator and what effect this has on the bias of the estimator.

9.2   The periodogram

Let $\{x(t), t = 0, 1, 2, \ldots n-1\}$ be a sequence of real-valued data,[6] and assume it comes from a stationary process $\{X_n, n \in \mathbb{Z}\}$ with mean zero, covariance function $r_X(\tau)$, and with continuous spectrum with density $R_X(f)$. Our goal is to estimate the spectral density of the process by means of the "spectrum" of the data sequence.

Many of the concepts defined in this chapter are applicable to any data sequence, regardless of whether it is a realization of a stationary process or not. Of course, when statistical properties such as expectation, variance, and correlation are involved, the stochastic process background is implicit.

9.2.1   Definition

Using the Fourier transform of the data vector,

$$\mathscr{X}(f) = \sum_{t=0}^{n-1} x(t) e^{-i2\pi ft}, \tag{9.1}$$

the *periodogram*, defined as

$$\widehat{R}_x(f) = \frac{1}{n} |\mathscr{X}(f)|^2, \tag{9.2}$$

is an estimate of the spectral density for the process. Obviously, the periodogram can be computed for any frequency f, not only for $-1/2 < f \leq 1/2$, as we are used to in the spectral density $R_X(f)$, and it is periodic, with period one.

---

[6]We have deliberately restricted ourselves to real-valued data, even if many of the methods are applicable also on complex valued data. In fact, many signal processing problems are naturally formulated in complex terms.

To bring it into a familiar form, we reformulate (9.2) as

$$\widehat{R}_x(f) = \frac{1}{n}\mathscr{X}^*(f) \cdot \overline{\mathscr{X}}(f) = \frac{1}{n}\sum_{t=0}^{n-1}\sum_{s=0}^{n-1} x(t)x(s)e^{-i2\pi f(s-t)}, \qquad (9.3)$$

and make a partial summation, collecting all terms $x(t)x(s)$ with the same lag difference $\tau = s - t$. For example, for $\tau = 0$, $t = s$ we have n possible combinations, from $x(0)x(0)$, up to $x(n-1)x(n-1)$, for $\tau = 1$, there are $n - 1$ combinations, $x(0)x(1), \ldots, x(n-2)x(n-1)$, and so on for larger values of τ. Thus, we get

$$\widehat{R}_x(f) = \sum_{\tau=-n+1}^{n-1}\frac{1}{n}\sum_{t=0}^{n-1-|\tau|} x(t)x(t+|\tau|)e^{-i2\pi f\tau} = \sum_{\tau=-n+1}^{n-1}\widehat{r}_x(\tau)e^{-i2\pi f\tau}, \quad (9.4)$$

and we recognize the estimate of the covariance function from (2.16),

$$\widehat{r}_x(\tau) = \frac{1}{n}\sum_{t=0}^{n-1-|\tau|} x(t)x(t+|\tau|), \qquad (9.5)$$

assuming $m = 0$.

In this chapter we will analyze the statistical properties of $\widehat{R}_x(f)$ as an estimate of $R_X(f)$, in particular its statistical distribution, expected value, and variance. We will show that it is asymptotically unbiased when $n \to \infty$, but that its variance does not go to zero, but stays (approximately) constant, $V[\widehat{R}_x(f)] \approx R_X^2(f)$ for $0 < |f| < 1/2$. We will also claim, but not prove, that for large n, the distribution of the periodogram estimate is

$$\frac{\widehat{R}_x(f)}{R_X(f)} \approx \frac{\chi^2(2)}{2}, \quad 0 < f < 1/2,$$

where $\chi^2(2)$ is a chi-square variable with two degrees of freedom.[7]

The mean square error of the spectrum estimate is

$$E[(\widehat{R}_x(f) - R_X(f))^2] = B_x^2(f) + V[\widehat{R}_x(f)], \qquad (9.6)$$

with squared bias

$$B_x^2(f) = (E[\widehat{R}_x(f)] - R_X(f))^2,$$

and variance

$$V[\widehat{R}_x(f)] = E[\widehat{R}_x(f)^2] - E^2[\widehat{R}_x(f)].$$

As the mean square error is so strongly affected by both the bias and the variance of the periodogram estimate, methods to reduce bias as well as variance become important.

---

[7] Note the difference between the Greek letter χ and the calligraphic $\mathscr{X}$ used for the Fourier transform of data x.

9.2.2   *Expected value*

The expected value of the periodogram is obtained by direct calculation as

$$E[\widehat{R}_x(f)] = \sum_{\tau=-n+1}^{n-1} E[\widehat{r}_x(\tau)]e^{-i2\pi f\tau} = \sum_{\tau=-n+1}^{n-1} \left(1 - \frac{|\tau|}{n}\right) r_x(\tau)e^{-i2\pi f\tau}, \quad (9.7)$$

where we have used the expression for $\widehat{r}_x(\tau)$ in (9.5) to obtain the second equality. When $n \to \infty$, it tends to

$$\sum_{\tau=-\infty}^{\infty} r_x(\tau)e^{-i2\pi f\tau} = R_X(f), \quad (9.8)$$

so the periodogram is an asymptotically unbiased estimate of the spectral density.

For small values of n, however, we rewrite

$$E[\widehat{R}_x(f)] = \sum_{\tau=-\infty}^{\infty} k_n(\tau)r_x(\tau)e^{-i2\pi f\tau}, \quad (9.9)$$

where $k_n(\tau) = \max(0, 1 - |\tau|/n)$ is called a *lag window*.

In the frequency plane, the multiplication transforms to a convolution between the spectral density and the Fourier transform of $k_n(\tau)$, as

$$
\begin{aligned}
E[\widehat{R}_x(f)] &= \sum_{\tau=-\infty}^{\infty} k_n(\tau) \int_{-1/2}^{1/2} R_X(u)e^{i2\pi u\tau}\, du\, e^{-i2\pi f\tau} \\
&= \int_{-1/2}^{1/2} R_X(u) \sum_{\tau=-\infty}^{\infty} k_n(\tau)e^{-i2\pi(f-u)\tau}\, du \\
&= \int_{-1/2}^{1/2} R_X(u)K_n(f-u)\, du, \quad (9.10)
\end{aligned}
$$

where $K_n(f) = \sum_{\tau=-n+1}^{n-1} k_n(\tau)e^{-i2\pi f\tau}$ is the Fourier transform of the lag window $k_n(\tau)$, known in the literature as the *Fejér kernel*.

Using the definition of $k_n(\tau)$, $K_n(f)$ can be found as

$$K_n(f) = nD_n^2(f) = \frac{\sin^2(n\pi f)}{n\sin^2(\pi f)},$$

where

$$D_n(f) = \frac{\sin(n\pi f)}{n\sin(\pi f)}$$

is the *Dirichlet kernel* . Figure 9.1 shows the typical shape of $K_n(f)$ in decibel (dB) scale.

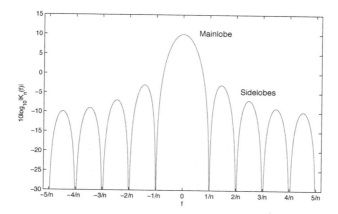

Figure 9.1 *The Fejér kernel,* $K_n(f)$.

Example 9.1. We illustrate the bias of the periodogram based on n observations of an AR(4)-process with poles located as $z_{1,2} = 0.95e^{\pm i2\pi 0.15}$ and $z_{3,4} = 0.95e^{\pm i2\pi 0.2}$.

In Figure 9.2, the true spectral density, $R_X(f)$, is plotted together with the expected value, $E[\widehat{R}_x(f)]$, of the periodogram for different values of n. For $n = 16$, the two peaks are not resolved, but for $n = 32$ and 128, the conclusion is that the spectrum consists of two peaks. For $f > 0.25$, however, there is severe bias for all values of n. ▲

Spectrum bias, $B_x(f) = E[\widehat{R}_x(f)] - R_X(f)$, is caused by the fact that an infinite length sequence in (9.8) is reduced to a finite length sequence in (9.9). This can be seen as a windowing of an infinite sequence of data, using a rectangle window of length n, which sets all other data samples equal to zero. The multiplication of the rectangular window function and the data transforms to a convolution in the frequency plane of the true spectrum (for the infinite sequence) and the kernel of the window function. The width of the so-called *mainlobe* of $K_n(f)$ is $2/n$, which becomes narrower with increasing n. The relative heights of the *sidelobes* remain at the same level independent of n, Figure 9.1. The bias of the peaks is caused by the mainlobe width and, consequently, the bias of the peaks decreases with increasing n. However, the bias for $f > 0.25$ is caused by the height of the sidelobes where power *leaks* from frequencies where the power is high ($f < 0.25$ in this case). This effect remains even if n is increased significantly and thereby this bias decreases very slowly, even for very large n.

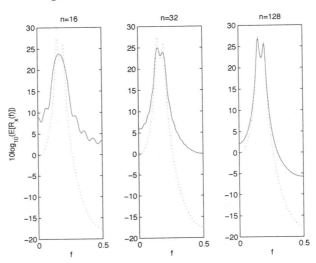

Figure 9.2 *Expected value,* $E[\widehat{R}_x(f)]$, *of the spectrum estimate for different data lengths n (solid lines). The dotted curve is the true spectral density* $R_X(f)$ *for the underlying AR(4)-process.*

9.2.3 *"Pre-whitening"*

The spectral density in the example had a large dynamic range, with narrow dominating peaks that "leak" over to adjacent frequencies in the estimate. For processes with an almost flat spectral density, the leakage is less serious. A common trick in practical frequency analysis is therefore to pre-whiten data. This is a technique to filter the data by a linear filter chosen so that the output has a flat, or almost flat, spectral density.

Pre-whitening can be achieved by fitting a parametric auto-regressive model (AR-process) to the data, by means of the estimation procedure in Section 7.3. For a true AR-process, filtering with the correct coefficients in (7.1) will produce uncorrelated output variables, i.e., white noise with constant spectral density. If the coefficients are not exactly correct, or the data do not come from an AR-process at all, one can anyway hope that the fitted AR-model produces residuals, $\widehat{e}_t = x_t - \sum_{k=1}^{p} \widehat{a}_k x_{t-k}$, that have a rather flat spectrum, if the order p is chosen high enough.

The spectral density of the whitened series is then estimated by the non-parametric methods in this chapter, and the result is transformed back to the original series with the help of (7.4).

9.2.4   *Variance*

To calculate the variance of the periodogram, we need to specify the distribution of the underlying process, and we assume it to be a Gaussian process with mean zero, giving

$$V[\widehat{R}_x(f)] = \frac{1}{n^2}C[|\mathscr{X}(f)|^2, |\mathscr{X}(f)|^2]$$
$$= \frac{1}{n^2}C[\mathscr{X}(f)\overline{\mathscr{X}}(f), \mathscr{X}(f)\overline{\mathscr{X}}(f)]. \qquad (9.11)$$

The Fourier transform, $\mathscr{X}(f)$, of a Gaussian real valued sequence $\{X(t), t = 0, 1, 2, \ldots n-1\}$ with zero mean, will also be a zero mean but complex valued Gaussian sequence, where its real and imaginary parts are real valued normal variables. The expectation of a complex random variable is $m_Z = E[Z] = E[\Re Z] + iE[\Im Z]$, and the covariance between Z_1 and Z_2 is defined as $C[Z_1, Z_2] = E[Z_1\overline{Z_2}] - E[Z_1]E[\overline{Z_2}]$. Hence, $V[Z] = C[Z, Z] = E[|Z - m_Z|^2] \geq 0$; see also Section A.2.2 in Appendix A.

The resulting variance for the periodogram can then be found by means of Isserlis' theorem, which we used on page 51 for real valued variables. For complex valued Gaussian random variables with zero mean it states that

$$C[Z_1 Z_2, Z_3 Z_4] = E[Z_1 Z_2 \overline{Z_3 Z_4}] - E[Z_1 Z_2]E[\overline{Z_3 Z_4}], \qquad (9.12)$$

where

$$E[Z_1 Z_2 \overline{Z_3 Z_4}] = E[Z_1 Z_2]E[\overline{Z_3 Z_4}] + E[Z_1 \overline{Z_3}]E[Z_2 \overline{Z_4}] + E[Z_1 \overline{Z_4}]E[Z_2 \overline{Z_3}]. \qquad (9.13)$$

Inserting (9.13) into (9.12) gives

$$C[Z_1 Z_2, Z_3 Z_4] = E[Z_1 \overline{Z_3}]E[Z_2 \overline{Z_4}] + E[Z_1 \overline{Z_4}]E[Z_2 \overline{Z_3}], \qquad (9.14)$$

as the last term of (9.12) is cancelled by the first term of (9.13). Noting that for zero mean random variables $C[Z_1, Z_3] = E[Z_1 \overline{Z_3}]$ as $E[Z_1] = E[Z_2] = 0$, we get the covariance

$$C[Z_1 Z_2, Z_3 Z_4] = C[Z_1, Z_3]C[Z_2, Z_4] + C[Z_1, Z_4]C[Z_2, Z_3]. \qquad (9.15)$$

The result is, with $Z_1 = Z_3 = \mathscr{X}(f), Z_2 = Z_4 = \overline{\mathscr{X}}(f)$,

$$V[\widehat{R}_x(f)] = \frac{1}{n^2}C[\mathscr{X}(f), \mathscr{X}(f)]C[\overline{\mathscr{X}}(f), \overline{\mathscr{X}}(f)])$$
$$+ \frac{1}{n^2}C[\mathscr{X}(f), \overline{\mathscr{X}}(f)]C[\overline{\mathscr{X}}(f), \mathscr{X}(f)]. \qquad (9.16)$$

For the different covariances (including $1/n$), we find

$$
\frac{1}{n}C[\mathscr{X}(f),\mathscr{X}(f)] = \frac{1}{n}C[\overline{\mathscr{X}}(f),\overline{\mathscr{X}}(f)])
$$
$$
= \frac{1}{n}E\left[\sum_{t=0}^{n-1}\sum_{s=0}^{n-1}X(t)X(s)e^{-i2\pi f(s-t)}\right] = R_X(f), \quad (9.17)
$$

and it can be shown that

$$
\frac{1}{n}C[\mathscr{X}(f),\overline{\mathscr{X}}(f)] = \frac{1}{n}C[\overline{\mathscr{X}}(f),\mathscr{X}(f),]
$$
$$
= \frac{1}{n}E\left[\sum_{t=0}^{n-1}\sum_{s=0}^{n-1}X(t)X(s)e^{-i2\pi f(s+t)}\right] \approx 0, \quad (9.18)
$$

for all frequencies not close to $f \neq [0,\pm 1/2]$. When $f = 0$ or $\pm 1/2$,

$$
\frac{1}{n}C[\mathscr{X}(f),\overline{\mathscr{X}(f)}] = \frac{1}{n}C[\overline{\mathscr{X}}(f),\mathscr{X}(f)] = R_X(f).
$$

We get

$$
V[\widehat{R}_x(f)] \approx \begin{cases} R_X^2(f), & \text{for } 0 < |f| < 1/2, \\ 2R_X^2(f), & \text{for } f = 0 \text{ and } \pm 1/2, \end{cases} \quad (9.19)
$$

asymptotically when $n \to \infty$.

Thus, the periodogram is inconsistent as an estimator of the spectral density since the variance does not decrease for large values of n.

9.3 The discrete Fourier transform and the FFT

The Discrete Fourier Transform (DFT) is necessary to allow the use of a computer to evaluate the Fourier transform for some distinct (discrete) frequency values. Calculation of discrete Fourier transforms is efficiently performed by the Fast Fourier Transform (FFT) algorithm, described in 1965 by Cooley and Tukey [15]. The method explores the structure of the Discrete Fourier Transform to minimize the number of multiplications and summations, and with the advent of powerful digital computers, the potential of the work could be fully exploited. The FFT has been reinvented many times during the years and today the original invention of the FFT algorithm is attributed to Carl Friedrich Gauss (1777-1855), in an unpublished work that was not printed until 1866.[8]

---

[8] C. F. Gauss, Nachlass: Theoria interpolationis methodo nova tractata, pp.265-303, in Carl Friedrich Gauss, Werke, Band 3, Göttingen: Königlichen Gesellschaft der Wissenschaften, 1866.

Let $\{x(t), t = 0, 1, 2, \ldots, n-1\}$ be a sequence of real valued data samples. To compute the DFT of these samples one can form

$$Z(k/n) = \sum_{t=0}^{n-1} x(t)e^{-i2\pi\frac{k}{n}t}, \quad k = 0, 1, \ldots, n-1, \qquad (9.20)$$

where $Z(k/n) = \mathscr{X}(f)$, when $f = k/n$.[9] For these frequencies the periodogram is defined as

$$\widehat{R}_x(k/n) = \frac{1}{n}|Z(k/n)|^2.$$

The Fourier transform (9.1) of a sequence of data is periodic with period one and thereby the DFT can be extended to a periodic function. Computing the DFT for $k = -k_0$ gives the same value as for $k = n - k_0$, meaning that when we study the values $0.5 < k/n < 1$, we see the spectrum that is also located at $-0.5 < -k/n < 0$. Computing the discrete Fourier transform of n samples using (9.20) will require about n^2 floating-point operations. If the length of the data sequence, n, happens to be a power of two, i.e., $n = 2^m$ for some integer m, the structure of the FFT allows $\widehat{R}_x(k/n)$ to be evaluated in merely $n\log_2 n$ floating-point operations, thus offering a substantial computational gain, especially for large data lengths. (This is easily checked with two test cases, comparing two cases $n_1 = 2^m$ and $n_2 = 2^m + 1$, where the last one would take considerably more time, at least for high values of m.)

However, likely, the data length n is not a power of two. In such cases, one may add $N - n$ zeros at the end of the sequence, such that N is a power of two. The additional zeroes will result in the spectral estimate being evaluated in N frequency grid points instead of only n, but will obviously not yield more information in the resulting estimate (although it may well appear so as the finer grid may visualize the spectral estimate better). The procedure to add zeros at the end of the sequence prior to computing the FFT is termed *zero-padding* and is a standard procedure in most computer programs.[10] Commonly, for small data lengths, we wish to compute the Fourier transform at additional frequency values in any case. The following example illustrates this.

Example 9.2. A discrete-time sinusoid with frequency $f = 0.125$ is calculated as $x(t) = \sin(2\pi f_0 t), t = 0, \ldots, n-1$, where $n = 16$. The periodogram is

---

[9]The standard definition of the DFT indexes the transform by k: $Z(k/n) = X_k = \sum_{t=0}^{n-1} x(t)e^{-i2\pi\frac{k}{n}t}$.

[10]In Section 5.6 we used zero-padding of a spectral density to increase the time resolution in a Monte Carlo simulation. Here we zero-pad the data to increase frequency resolution.

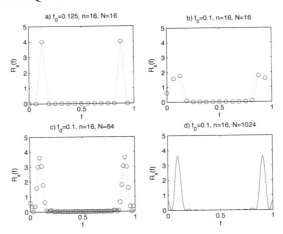

Figure 9.3 *The periodogram of* $x(t) = \sin(2\pi f_0 t)$, $t = 0, \ldots, 15$, *using FFT with N values.*

computed for $n = 16$ frequency values and the result is seen in Figure 9.3(a), with a clear peak at the frequency $f_0 = 0.125$ and another one at 0.875, which we recall is the same as having a peak at $0.875 - 1 = -0.125$, in accordance with the Fourier transform of a real-valued sinusoid which becomes a delta function; see Section 4.3.2. All other values are zero, which is what to expect.

If the frequency of the sinusoid is slightly changed, e.g., to $f_0 = 0.1$, the view of the periodogram changes to Figure 9.3(b), where instead the spectrum estimate at several frequency values differ from zero and no clear peak is seen at $f_0 = 0.1$. To actually be able to interpret the periodogram, the calculation has to be made using more closely spaced frequency values. Using zero-padding, see Figure 9.3(c,d), where $N = 64$ and $N = 1024$ are used, the actual spectrum estimate becomes more visible. The true spectral density of a sinusoid of infinite length is a delta function but as the sinusoid is of finite length, $n = 16$, the spectrum becomes the delta function convolved with the Fejér kernel function $K_n(f)$; see (9.10).

The sidelobe leakage is seen at nearby frequencies and especially at $f = 0$, where the sidelobes of the sinusoid component at $f_0 = 0.1$ and $f_0 = -0.1$ interfere and thereby the spectrum at $f = 0$ becomes quite large. The mainlobe width will be more narrow for larger values of n but still the leakage from the sidelobes will influence nearby frequencies and in the case of several closely spaced sinusoids, especially if they have different power, a small power component might be hidden in the sidelobe leakage of a stronger one. ▲

As discussed above, we often use zero-padding up to a total length that fulfills $N = 2^m$ and then the FFT algorithms can be used for faster calculation. The computational cost of using much more zero-padding, e.g., $N = 1024$, instead of $N = 64$ is small. The view of the spectrum estimate becomes much clearer and easier to interpret; see Figure 9.3(d). It should be noted that in Figure 9.3(a), the $n = 16$ frequency samples were taken at the exact locations where the Fejér kernel is zero (except for the sinusoid frequency $f = 0.125$) and thereby the view of the 16-sample sinusoid became as if N would have been infinite. Computing the spectrum estimate for a larger value of N will show the true picture, where the Fejér kernel has been convolved with the delta function at $f_0 = 0.125$ (and $f_0 = -0.125$).

9.4   Bias reduction – data windowing

The periodogram has a severe drawback in the bias caused by the sidelobes of the kernel $K_n(f)$. If the sidelobes were lower, some of the bias caused by leakage in the estimate could be reduced, which is especially important when the dynamical range of the spectrum is large; see, e.g., Figure 9.2. One solution to this is to use *windowing* or *tapering* of the data, where equations (9.1) and (9.2) are extended to

$$\widehat{R}_w(f) = \frac{1}{n} \left| \sum_{t=0}^{n-1} x(t) w(t) e^{-i2\pi ft} \right|^2, \tag{9.21}$$

which sometimes is called the *modified periodogram*. Expressing the data samples in the inverse Fourier transform of data, $x(t) = \int_{-1/2}^{1/2} \mathscr{X}(v) e^{i2\pi vt} \, dv$, the modified periodogram could also be rewritten as

$$\widehat{R}_w(f) = \frac{1}{n} \left| \sum_{t=0}^{n-1} \int_{-1/2}^{1/2} \mathscr{X}(v) e^{i2\pi vt} \, dv \, w(t) e^{-i2\pi ft} \right|^2$$

$$= \frac{1}{n} \left| \int_{-1/2}^{1/2} \mathscr{X}(v) \sum_{t=0}^{n-1} w(t) e^{-i2\pi(f-v)t} \, dv \right|^2$$

$$= \frac{1}{n} \left| \int_{-1/2}^{1/2} \mathscr{X}(v) W(f-v) \, dv \right|^2, \tag{9.22}$$

where $W(f)$ is the Fourier transform of the window function $w(t)$, $W(f) = \sum_{t=0}^{n-1} w(t) e^{-i2\pi ft}$.

There is a huge literature on data windowing but here we will just mention a few examples. One of the most commonly used windows is the Hann or

Hanning window,

$$h(t) = \frac{1}{2} - \frac{1}{2}\cos\left(\frac{2\pi t}{n-1}\right), \, t = 0,\ldots,n-1, \qquad (9.23)$$

which often is normalized according to

$$w(t) = \frac{h(t)}{\sqrt{\frac{1}{n}\sum_{t=0}^{n-1}h^2(t)}}, \qquad (9.24)$$

for unchanged estimated power. The window function and the corresponding window spectrum,

$$K_w(f) = \frac{1}{n}\left|\sum_{t=0}^{n-1}w(t)e^{-i2\pi ft}\right|^2 = |W(f)|^2,$$

is seen in Figure 9.4(a) and (b) for $n = 32$ (solid lines). For comparison, the shape of the rectangle window and window spectrum is also depicted (dotted lines). The Hanning window has a wider mainlobe than the rectangle window ($4/n$ compared to $2/n$), but the sidelobes decrease much faster which prevents leakage.

The difference in the expected value of the spectrum estimate

$$E[\widehat{R}_w(f)] = \int_{-1/2}^{1/2} R_X(u)K_w(f-u)\,du \qquad (9.25)$$

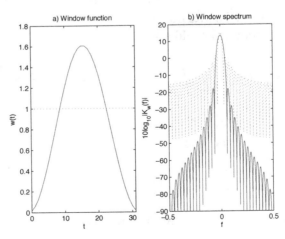

Figure 9.4 *(a) The Hanning window function (solid line) and the rectangle window function (dotted line) for* $n = 32$; *(b) the corresponding Hanning window spectrum (solid line) and rectangle window spectrum, the Fejér kernel function (dotted line).*

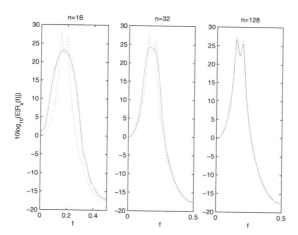

Figure 9.5 *The expected value of the spectrum estimate using the Hanning window for different data length n.*

is clearly seen if we compare the results of Figure 9.5, where Hanning windows of different length have been used, with the results of Figure 9.2. For the length $n = 32$, the two peaks are not resolved for the Hanning window. The reason is the doubled width of the mainlobe of the Hanning window. However, for $f > 0.25$, the spectrum is estimated with much smaller bias for the Hanning window.

Usually the data length is limited and we must be careful in our choice of window. Typically, the choice depends on the characteristics of the measured signal. The rectangle window (the Fejér kernel) has the most narrow mainlobe of all windows $(2/n)$, and should be the choice if the dynamic range (power variation) of the spectrum is small. In the case of a high dynamic range, the window with lowest possible sidelobes is the Slepian window or the zeroth-order *discrete prolate spheroidal sequence*, (DPSS), [42], depicted in Figure 9.6(a) and (b). The Slepian window has the most concentrated power inside a predefined bandwidth B defining the mainlobe width, i.e.,

$$\max_{K_w(f)} \frac{\int_{-B/2}^{B/2} K_w(f)\,\mathrm{d}f}{\int_{-1/2}^{1/2} K_w(f)\,\mathrm{d}f}.$$

Consequently, the power of the sidelobes will be very small. The advantage of this window is clearly seen in an example where two sinusoids differ a lot in power; the smaller one might be hidden by the sidelobes of the stronger one.

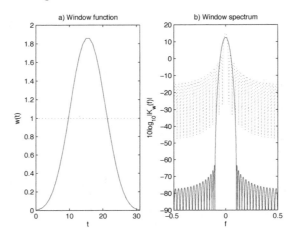

Figure 9.6 *(a) The Slepian window function (solid line) and the rectangle window function (dotted line) for $n = 32$; (b) the corresponding Slepian window spectrum (solid line) and rectangle window spectrum (dotted line).*

Example 9.3. The spectrum of a sum of two sinusoids,

$$x(t) = \cos 2\pi 0.1 t + 0.0001 \cos 2\pi 0.3 t, \quad t = 0 \ldots 99,$$

is calculated using the rectangular window, the Hanning window, and the Slepian window. We show an example of the resulting spectra in Figure 9.7. It is not possible to identify the peak with the smaller power located at $f = 0.3$ using the rectangle window. The sidelobe leakage from the frequency with stronger power at $f = 0.1$ is too large and is hiding the smaller power of the sinusoid at $f = 0.3$. With use of the Hanning window, the peak at $f = 0.3$ is seen but might be misinterpreted in a more noisy situation. Using a Slepian window with $B = 0.08$, both peaks are easily seen as the sidelobe leakage of the Slepian window is very low. The disadvantage, however, is the increased width of the sinusoidal peaks. Compare the shapes of the window spectra in Figures 9.4 and 9.6. ▲

The variance of the modified periodogram for $n \to \infty$ is approximately equal to the variance for the periodogram, i.e.,

$$\mathsf{V}[\widehat{R}_w(f)] \approx \begin{cases} R_X^2(f), & \text{for } 0 < |f| < 1/2, \\ 2R_X^2(f), & \text{for } f = 0 \text{ and } \pm 1/2. \end{cases} \tag{9.26}$$

For the periodogram, if we restrict ourselves to the Fourier frequencies, $f_k = k/n, n = 0, \ldots, n/2$, the pairwise covariance

$$\mathsf{C}[\widehat{R}_x(f_k), \widehat{R}_x(f_l)] \approx 0, \quad k \neq l, \tag{9.27}$$

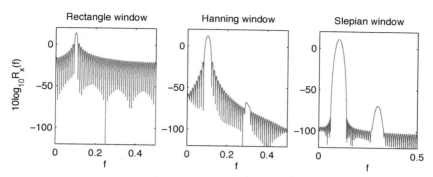

Figure 9.7 *Leakage properties of different windows for two sinusoids with different power using different window functions, $n = 100$.*

if n is large enough. For the modified periodogram (9.27), this is not necessarily true for $f_k = k/n$. Instead, we find that

$$C[\widehat{R}_w(f_k), \widehat{R}_w(f_l)] = \frac{1}{n^2} C[\mathscr{X}_w(f_k)\overline{\mathscr{X}_w}(f_k), \mathscr{X}_w(f_l)\overline{\mathscr{X}_w}(f_l))]$$

$$\approx \frac{1}{n^2} |E[\overline{\mathscr{X}_w}(f_k)\mathscr{X}_w(f_l)]|^2, \qquad (9.28)$$

using equations (9.16) and (9.18) and where

$$\mathscr{X}_w(f) = \int_{-1/2}^{1/2} \mathscr{X}(v)W(f - v)\,dv.$$

An approximate expression for the covariance is then

$$C[\widehat{R}_w(f_k), \widehat{R}_w(f_l)] \approx \frac{1}{n^2} \left| \int_{-1/2}^{1/2} \overline{W}(f_k - v)W(f_l - v)R_X(v)\,dv \right|^2$$

$$\leq \frac{\max R_X(f)^2}{n^2} \left| \int_{-1/2}^{1/2} \overline{W}(f_k - v)W(f_l - v)\,dv \right|^2,$$

and we conclude that the spectral width of the window function is highly influencing the covariance between frequency values.

9.5 Reduction of variance

A central question in frequency analysis of stationary processes is how to produce a consistent estimate of the spectral density. That requires a reduction of variance with increasing sample size. We shall describe several methods to decrease variance and see how the bias is affected.

9.5.1   Lag windowing

In view of (9.27), a natural way of reducing the variance is to smooth across nearby frequencies, i.e.,

$$\widehat{R}_{smooth}(f_k) = \frac{1}{2L+1} \sum_{j=-L}^{L} \widehat{R}_x(f_{k-j}), \tag{9.29}$$

to obtain a reduction of the variance. The bias will not be much affected, since $E[\widehat{R}_x(f_{k-j})] \approx E[\widehat{R}_x(f_k)]$, $j = -L, \dots, L$, for any smooth spectral density $R_X(f)$. Another way of expressing equation (9.29) is

$$\widehat{R}_{lw}(f) = \int_{-1/2}^{1/2} K_{L_n}(f - v)\widehat{R}_x(v)\,dv, \tag{9.30}$$

where $K_{L_n}(f)$ is the smoothing function across frequencies. We could rewrite this as

$$\widehat{R}_{lw}(f) = \sum_{\tau=-\infty}^{\infty} k_{L_n}(\tau)\widehat{r}_x(\tau)e^{-i2\pi f\tau},$$

where the *lag window* $k_{L_n}(\tau)$, $-L_n \leq \tau \leq L_n$, $(L_n < n)$, is used to weight the different samples of the estimated covariance function. The parameter L_n is often chosen much smaller than n as the covariance estimates for large values of τ are based on very few data samples (the last one, $\tau = n - 1$, is actually just based on $x(0)x(n-1)$), and these estimates increase the variance of the spectrum estimate. This was first suggested by Blackman and Tukey, [6], and the reason for using the lag window is that the variance of $\widehat{R}_x(f)$ does not decrease even when n is large. By defining an appropriate lag window that reduces the effect of the most unreliable covariance estimates for large $|\tau|$, the variance is decreased.

We can get an approximate variance expression as

$$V[\widehat{R}_{lw}(f)] \approx \frac{R_X^2(f)}{B_w n},$$

where the bandwidth B_w of the spectral estimator is a measure of the minimum separation in frequency between approximately uncorrelated spectral estimates and is computed as

$$B_w = \frac{(\int_{-1/2}^{1/2} K_{L_n}(f)\,df)^2}{\int_{-1/2}^{1/2} K_{L_n}^2(f)\,df}. \tag{9.31}$$

With the normalization

$$\int_{-1/2}^{1/2} K_{L_n}(f)\,df = k_{L_n}(0) = 1,$$

Equation (9.31) simplifies to

$$B_w = \frac{1}{\int_{-1/2}^{1/2} K_{L_n}^2(f)\,df} = \frac{1}{\sum_{\tau=-L_n}^{L_n} k_{L_n}^2(\tau)}.$$

9.5.2 Averaging of spectra

Other ways of dealing with the variance aspect have been proposed. Bartlett was the first to suggest to reduce the variance of the spectrum estimate by averaging of several uncorrelated periodograms. He was followed by Welch who invented the WOSA (Weighted Overlap Segmented Averaging) algorithm, which is frequently used in many application areas, [52]. If we divide the data in shorter sequences, i.e., n samples of data are divided in K sequences with approximately n/K samples each, and then calculate the average of the K different spectrum estimates, $\widehat{R}_{x,k}(f), k = 1\ldots K,$

$$\widehat{R}_{av}(f) = \frac{1}{K} \sum_{k=1}^{K} \widehat{R}_{x,k}(f), \qquad (9.32)$$

a reduction of the variance is achieved if the spectrum estimates $\widehat{R}_{x,k}(f)$ and $\widehat{R}_{x,l}(f)$ have small correlation for $k \neq l$. In this case, the variance is

$$V[\widehat{R}_{av}(f)] = \frac{1}{K^2} \sum_{k=1}^{K} \sum_{l=1}^{K} C[\widehat{R}_{x,k}(f), \widehat{R}_{x,l}(f)] = \frac{1}{K^2} \sum_{k=1}^{K} V[\widehat{R}_{x,k}(f)]$$

$$\approx \frac{1}{K} R_X^2(f), \qquad (9.33)$$

using the approximation from (9.19), for large values of n.

Note that the bias now is

$$E[\widehat{R}_{av}(f)] = \frac{1}{K} \sum_{k=1}^{K} \int_{-1/2}^{1/2} R_{X,k}(u) K_{n/K}(f-u)\,du = \int_{-1/2}^{1/2} R_X(u) K_{n/K}(f-u)\,du,$$

which has increased considerably as the kernel mainlobe width has become a factor K wider since the windowed sequence is of length n/K samples instead of n samples. If the process X_t has uncorrelated samples (i.e., is a white noise sequence), the Fourier transforms of the non-overlapping parts of the sequence are also uncorrelated and (9.33) is valid. When the process samples are correlated, e.g., for an AR-process, this correlation will transfer to the Fourier transforms of the parts. However, (9.33) will still be approximately true and a considerable reduction of the variance will be achieved.

The distribution of the averaged periodogram estimate is

$$\frac{\widehat{R}_{av}(f)}{R_X(f)} \approx \frac{\chi^2(2K)}{2K}, \quad 0 < f < 1/2,$$

where $\chi^2(2K)$ is a chi-square random variable with $2K$ degrees of freedom. For large values of K and n, it holds approximatively that

$$\frac{\widehat{R}_{av}(f)}{R_X(f)} \sim N(1, 1/K).$$

Example 9.4. To show the performance, we illustrate with an example where ten realizations from an AR(4)-process of length $n = 128$, with poles in $0.95e^{\pm i2\pi 0.15}$ and $0.95e^{\pm i2\pi 0.27}$, are used. The periodogram estimates using the Hanning window of length 128 are calculated and plotted in Figure 9.8(a) as dotted lines. The variation of the ten different estimates (seen as the gray blurring) illustrates the large standard deviation which is approximately equal to $R_X(f)$.

Dividing each 128 sample sequence into $K = 8$ non-overlapping smaller pieces (each of length 16 samples) and averaging the periodograms (Hanning windowed) from the different (128/8) 16 samples sequences will give an estimate with variance reduction of about 8 times. This is illustrated in Figure 9.8(b) where the variation of the estimates of the same ten realizations is much smaller than in Figure 9.8(a). However, as the sequences of data are much shorter (16 samples), severe bias is introduced in the estimates. The view of the two peaks are no longer clear.

The trade-off between bias and variance becomes important and will of course be highly dependent of the application, the resolution, and dynamics of the actual spectrum. To achieve the best trade-off, overlap are often applied, using the Welsh WOSA-algorithm. It has been shown that a 50% overlap of the Hanning window often gives the best trade-off between bias and variance. We illustrate this in Figure 9.8(c) where the individual sequences from one realization now are of length 32 samples and chosen with overlap 50%, which gives $K = 7$ segments from the total data length of $n = 128$ samples. The ten estimates now all follow the true spectrum rather closely as the bias is reduced by the longer sequence length. The variance is reduced almost as much as in Figure 9.8(b) as the number of sequences is almost as large ($K = 7$ compared to $K = 8$). ▲

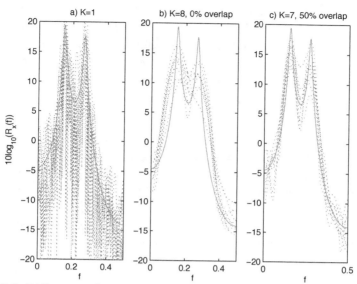

Figure 9.8 *Different realizations of data lengths $n = 128$ of an AR(4)-process are used to show the variance of the spectrum estimates. The estimation is carried out with 10 different realizations (dotted lines) and the true power density spectrum is depicted as the solid line; (a) the periodogram estimates; (b) Welch method with 0% overlap; (c) Welch method with 50% overlap.*

9.5.3 Multiple windows

The requirement for the averaging of periodograms to actually reduce variance is that the periodograms are almost uncorrelated. The drawback of this method is that just a small amount of the total data sequence is used for each periodogram. A better method would be if we could use the whole data length for all periodograms in the average, but the spectrum estimates should still be uncorrelated. Since the data sequences are the same, the only possibility to achieve uncorrelated spectra is to have different windows for the different periodograms. This is the concept of *Multiple windows* or *Multitapers*, which was invented by David Thomson, [46], giving the multiple window spectrum estimate as

$$\widehat{R}_x(f) = \frac{1}{K}\sum_{k=1}^{K}\widehat{R}_{x,k}(f) = \frac{1}{K}\sum_{k=1}^{K}\left|\sum_{t=0}^{n-1}x(t)h_k(t)e^{-i2\pi ft}\right|^2. \tag{9.34}$$

The question then is how should the windows $h_k(t)$, $t = 0,\ldots,n-1$, $k = 1,\ldots,K$, be chosen? Actually, to fulfill the requirement of uncorrelated periodograms, the set of multiple windows has to depend on the true, unknown, spectral density. The approach suggested by Thomson was that the estimates should be uncorrelated for a white noise spectrum, and thereby give the optimal variance reduction in this case. It is easily shown that the periodogram estimates will be uncorrelated if

$$\sum_{t=0}^{n-1} h_k(t)h_l(t) = 0, \quad l \neq k.$$

Resolution and leakage are other important properties and Thomson chose to construct the windows to give the smallest possible leakage in a predefined bandwidth B of the white noise. The resulting functions are shown to be the Slepian sequences, [42]. The bandwidth B is also the final resolution of the spectrum estimate and can be shown to be approximately connected to the window/sequence length n and number of windows K as

$$K \approx n \cdot B - 2.$$

This formula indicates that for a fixed n, if we would like to reduce the variance, i.e., increase K, the value of B must increase as well, which means that the resolution decreases. The example below illustrates this.

Example 9.5. An example with the same AR(4)-process as in Example 9.4 is shown in Figure 9.9, where the multiple window spectrum estimates of the same ten realizations are plotted. Different numbers of windows K, and thereby also different resolutions, are chosen to show the performance. Figure 9.9(a) shows the results of a smaller number of windows ($K = 3$), which gives a better resolution but a larger variance. In Figure 9.9(b) and (c), the bandwidths are increased to $B = 0.08$ and $B = 0.12$, respectively. A wider band decreases the resolution as it smoothes the spectrum in the band. The resulting estimates will show a box-like shape of a peak in the spectrum. In Figure 9.9(c), the resolution is so bad that the two peaks are no longer identifiable. With a proper choice of the parameters, both bias and variance of the estimate could be acceptable. However, it has also been shown that for varying spectra, such as peaked spectra, the Thomson multiple window method degrades due to cross-correlation between spectra, [49]. This is notable, e.g., in Figure 9.9(b) and (c) at the peak frequencies, where the variances are large in spite of the large number of windows. ▲

9.5.4 Estimation of cross- and coherence spectrum

We have seen examples in previous chapters where the cross-spectrum and coherence spectrum have been computed. Naturally, a first choice to evaluate

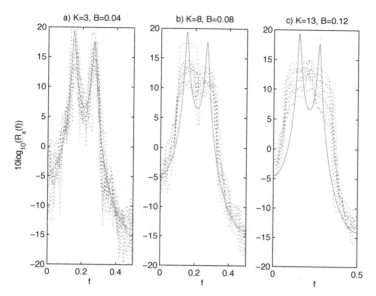

Figure 9.9 *Result of multiple window spectrum estimation of an AR(4)-process with different number of windows K. The estimation is carried out with 10 different realizations (dotted lines) and the true power density spectrum is depicted as the full line.*

such estimates could be based on the periodogram, (9.2). However, using the periodogram for the coherence spectrum produce a strange result. If we study the formulas for the coherence spectrum estimate using the periodogram (the formulas for the modified periodogram will be similar), we get

$$\widehat{\kappa_{x,y}^2} = \frac{|\widehat{R}_{x,y}(f)|^2}{\widehat{R}_x(f)\widehat{R}_y(f)} = \frac{\frac{1}{n}\mathscr{X}(f)\overline{\mathscr{Y}}(f)\cdot\frac{1}{n}\overline{\mathscr{X}}(f)\mathscr{Y}(f)}{\frac{1}{n}\mathscr{X}(f)\overline{\mathscr{X}}(f)\cdot\frac{1}{n}\mathscr{Y}(f)\overline{\mathscr{Y}}(f)} = 1. \tag{9.35}$$

The result is one for all frequencies, independent of what $\mathscr{X}(f)$ and $\mathscr{Y}(f)$ are. Thus, the periodogram and the modified periodogram are not useful estimators for the coherence spectrum.

For a more reliable coherence spectrum, the averaged spectrum of (9.32) could be used and (9.35) will change into

$$\widehat{\kappa_{x,y}^2} = \frac{|\frac{1}{K}\sum_{j=1}^K \widehat{R}_{xy,j}(f)|^2}{\frac{1}{K}\sum_{j=1}^K \widehat{R}_{x,j}(f)\frac{1}{K}\sum_{j=1}^K \widehat{R}_{y,j}(f)}$$

$$= \frac{\frac{1}{nK}\sum_{j=1}^K \mathscr{X}_j(f)\overline{\mathscr{Y}}_j(f)\cdot\frac{1}{nK}\sum_{j=1}^K \overline{\mathscr{X}}_j(f)\mathscr{Y}_j(f)}{\frac{1}{nK}\sum_{j=1}^K \mathscr{X}_j(f)\overline{\mathscr{X}}_j(f)\cdot\frac{1}{nK}\sum_{j=1}^K \mathscr{Y}_j(f)\overline{\mathscr{Y}}_j(f)}.$$

Example 9.6. We illustrate the coherence spectrum with an example of a respiratory variation measurement and the corresponding heart rate variability (HRV) signal ($n = 720$). Coherence spectra between these two signals are seen in Figure 9.10, where the number of windows are $K = 2$ (zero overlap), 4 (50% overlap), and 8 (75% overlap) and the Hanning window length for each periodogram is 256 in all cases. Studying the different estimates of the coherence spectral densities shows that it is important to be careful to find a reliable coherence estimate. In Figure 9.10(a), the coherence spectrum seems to be close to one, not only for the strong power frequency 0.22 Hz, but also for 0.12 Hz. With 50% overlap (Figure 9.10(b)), the number of windows increases to $K = 4$, which reduces the variances of the spectra approximately by a factor two, meaning that this coherence spectrum should be more reliable. In Figure 9.10(c), the number of windows is increased to $K = 8$, which also reduce the variances somewhat, although not as much as a factor 2, as the overlap of the windows is too large (75%). A more extensive view of the variance on the coherence spectrum can be found in [21].   ▲

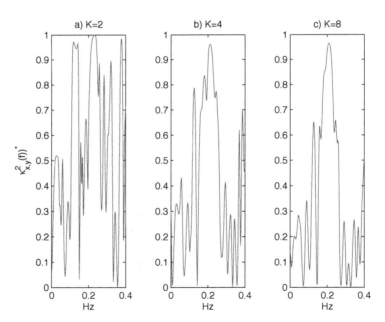

Figure 9.10 *The coherence spectrum of a respiratory signal and a HRV-signal for different number of averaged spectra; (a) K=2; (b) K=4; (c) K=8.*

Exercises

9:1. Show that the Fourier transform of $\{x(t), t = 0, 1, \ldots n - 1\}$,

$$\mathscr{X}(f) = \sum_{t=0}^{n-1} x(t) e^{-i2\pi ft},$$

included in the periodogram, $\widehat{R}_x(f) = \frac{1}{n} |\mathscr{X}(f)|^2$, is periodic with period one.

9:2. Show that the mean square error of the spectrum estimate is equal to the sum of the variance and the squared bias,

$$E[(\widehat{R}_x(f) - R_x(f))^2] = V[\widehat{R}_x(f)] + B_x^2(f),$$

where the variance defined as

$$V[\widehat{R}_x(f)] = E[(\widehat{R}_x(f))^2] - E^2[\widehat{R}_x(f)],$$

and the squared bias

$$B_x^2(f) = (E[\widehat{R}_x(f)] - R_x(f))^2.$$

9:3. Figure 9.11(a) shows the periodogram and Figure 9.11(b) shows the modified periodogram using a Hanning window of the same data sequence with $n = 64$ samples. We know that the sequence consists of a number of sinusoids. By studying the figures, what can you conclude about the sequence of data?

9:4.(a) Determine the condition of the data window function $w(t)$, for an unbiased spectral estimate of a white noise process, $x(t)$, $t = 0 \ldots n - 1$, with variance σ^2. Hint: The expected value can be written as

$$E[\widehat{R}_w(f)] = \int_{-1/2}^{1/2} R_x(u) K_w(f - u) du = \int_{-1/2}^{1/2} R_x(u) |W(f - u)|^2 du,$$

where $W(f)$ is the Fourier transform of $w(t)$.

(b) If we also consider the variance, how is the "best" spectrum estimate (lowest variance) for a white noise process achieved?

9:5. Figure 9.12(a) shows the modified periodogram and Figure 9.12(b) the Welch method of 10 realizations of an AR(2)-process with a peak at $f = 0.2$. The number of data samples is $n = 64$ and the number of averages in Welch method is $K = 8$. Which of the two methods gives the best estimate? Motivate your answer.

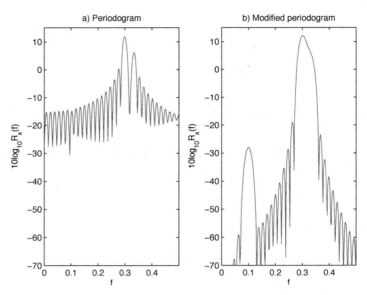

Figure 9.11 *Spectrum estimates for Exercise 9:3 using periodogram and modified periodogram of the same data sequence.*

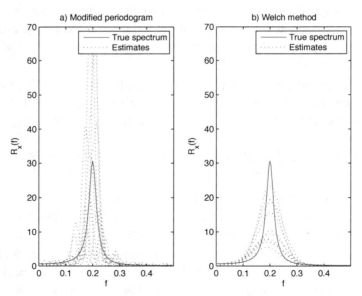

Figure 9.12 *Spectrum estimates to Exercise 9:5 using modified periodogram and the Welch method of the same data sequence.*

Appendix A

Some probability and statistics

A.1 Probabilities, random variables and their distribution

We summarize a few of the basic concepts of random variables, usually denoted by capital letters, X, Y, Z, etc, and their probability distributions, defined by the *cumulative distribution function* (CDF) $F_X(x) = P(X \leq x)$, etc.

To a random experiment we define a sample space Ω, that contains all the outcomes that can occur in the experiment. A random variable, X, is a function defined on a sample space Ω. To each outcome $\omega \in \Omega$ it defines a real number $X(\omega)$, that represents the value of a numeric quantity that can be measured in the experiment. For X to be called a random variable, the probability $P(X \leq x)$ has to be defined for all real x.

Distributions and moments

A probability distribution with cumulative distribution function $F_X(x)$ can be discrete or continuous with *probability function* $p_X(x)$ and *probability density function* $f_X(x)$, respectively, such that

$$F_X(x) = P(X \leq x) = \begin{cases} \sum_{k \leq x} p_X(k), & \text{if } X \text{ takes only integer values,} \\ \int_{-\infty}^{x} f_X(y) \, dy. \end{cases}$$

A distribution can be of *mixed type*. The distribution function is then an integral plus discrete jumps; see Appendix B.

The *expectation* of a random variable X is defined as the center of gravity in the distribution,

$$E[X] = m_X = \begin{cases} \sum_k k p_X(k), \\ \int_{x=-\infty}^{\infty} x f_X(x) \, dx. \end{cases}$$

The *variance* is a simple measure of the spreading of the distribution and is

261

defined as

$$V[X] = E[(X - m_X)^2] = E[X^2] - m_X^2 = \begin{cases} \sum_k (k - m_X)^2 p_X(k), \\ \int_{x=-\infty}^{\infty} (x - m_X)^2 f_X(x)\,dx. \end{cases}$$

Chebyshev's inequality states that, for all $\varepsilon > 0$,

$$P(|X - m_X| > \varepsilon) \leq \frac{E[(X - m_X)^2]}{\varepsilon^2}.$$

In order to describe the statistical properties of a random function one needs the notion of *multivariate distributions*. If the result of an experiment is described by two different quantities, denoted by X and Y, e.g., length and height of randomly chosen individual in a population, or the value of a random function at two different time points, one has to deal with a two-dimensional random variable. This is described by its two-dimensional distribution function $F_{X,Y}(x,y) = P(X \leq x, Y \leq y)$, or the corresponding two-dimensional probability or density function,

$$f_{X,Y}(x,y) = \frac{\partial^2 F_{X,Y}(x,y)}{\partial x \partial y}.$$

Two random variables X, Y are *independent* if, for all x, y,

$$F_{X,Y}(x,y) = F_X(x)F_Y(y).$$

An important concept is the *covariance* between two random variables X and Y, defined as

$$C[X,Y] = E[(X - m_X)(Y - m_Y)] = E[XY] - m_X m_Y.$$

The *correlation coefficient* is equal to the dimensionless, normalized covariance,[1]

$$\rho[X,Y] = \frac{C[X,Y]}{\sqrt{V[X]V[Y]}}.$$

Two random variables with zero correlation, $\rho[X,Y] = 0$, are called *uncorrelated*. Note that if two random variables X and Y are independent, then they are also uncorrelated, but the reverse does not hold. It can happen that two uncorrelated variables are dependent.

---

[1]Speaking about the "correlation between two random quantities," one often means the degree of covariation between the two. However, one has to remember that the correlation only measures the degree of *linear covariation*. Another meaning of the term correlation is used in connection with two data series, $(x_1, \ldots, x_n)$ and $(y_1, \ldots, y_n)$. Then sometimes the sum of products, $\sum_1^n x_k y_k$, can be called "correlation," and a device that produces this sum is called a "correlator."

Conditional distributions

If X and Y are two random variables with bivariate density function $f_{X,Y}(x,y)$, we can define the *conditional distribution* of X given $Y = y$, by the conditional density,

$$f_{X|Y=y}(x) = \frac{f_{X,Y}(x,y)}{f_Y(y)},$$

for every y where the *marginal density* $f_Y(y)$ is non-zero. The expectation in this distribution, the *conditional expectation*, is a function of y, and is denoted and defined as

$$E[X \mid Y = y] = \int_x x f_{X|Y=y}(x)\, dx = m(y).$$

The *conditional variance* is defined as

$$V[X \mid Y = y] = \int_x (x - m(y))^2 f_{X|Y=y}(x)\, dx = \sigma_{X|Y}(y).$$

The unconditional expectation of X can be obtained from the *law of total probability*, and computed as

$$E[X] = \int_y \left\{ \int_x x f_{X|Y=y}(x)\, dx \right\} f_Y(y)\, dy = E[E[X \mid Y]].$$

The unconditional variance of X is given by

$$V[X] = E[V[X \mid Y]] + V[E[X \mid Y]].$$

A.2   Multidimensional normal distribution

A one-dimensional normal random variable X with expectation m and variance σ^2 has probability density function

$$f_X(x) = \frac{1}{\sqrt{2\pi\sigma^2}} \exp\left\{ -\frac{1}{2}\left(\frac{x-m}{\sigma}\right)^2 \right\},$$

and we write $X \sim N(m, \sigma^2)$. If $m = 0$ and $\sigma = 1$, the normal distribution is standardized. If $X \sim N(0,1)$, then $\sigma X + m \in N(m, \sigma^2)$, and if $X \sim N(m, \sigma^2)$ then $(X - m)/\sigma \sim N(0,1)$. We accept a constant random variable, $X \equiv m$, as a normal variable, $X \sim N(m, 0)$.

Now, let $X_1, \ldots, X_n$ have expectation $m_k = E[X_k]$ and covariances $\sigma_{jk} = C[X_j, X_k]$, and define (with $'$ for transpose),

$$\boldsymbol{\mu} = (m_1, \ldots, m_n)',$$
$$\boldsymbol{\Sigma} = (\sigma_{jk}) = \text{the covariance matrix for } X_1, \ldots, X_n.$$

It is a characteristic property of the normal distributions that all linear combinations of a multivariate normal variable also have a normal distribution. To formulate the definition, write $\mathbf{a} = (a_1, \ldots, a_n)'$ and $\mathbf{X} = (X_1, \ldots, X_n)'$, with $\mathbf{a}'\mathbf{X} = a_1 X_1 + \cdots + a_n X_n$. Then,

$$
\begin{aligned}
E[\mathbf{a}'\mathbf{X}] = \mathbf{a}'\boldsymbol{\mu} &= \sum_{j=1}^{n} a_j m_j, \\
V[\mathbf{a}'\mathbf{X}] = \mathbf{a}'\boldsymbol{\Sigma}\mathbf{a} &= \sum_{j,k}^{n} a_j a_k \sigma_{jk}.
\end{aligned}
\tag{A.1}
$$

> **Definition A.1.** *The random variables* $X_1, \ldots, X_n$, *are said to have an n-dimensional normal distribution if every linear combination* $a_1 X_1 + \cdots + a_n X_n$ *has a normal distribution. From (A.1) we have that* $\mathbf{X} = (X_1, \ldots, X_n)'$ *is n-dimensional normal, if and only if* $\mathbf{a}'\mathbf{X} \sim N(\mathbf{a}'\boldsymbol{\mu}, \mathbf{a}'\boldsymbol{\Sigma}\mathbf{a})$ *for all* $\mathbf{a} = (a_1, \ldots, a_n)'$.

Obviously, $X_k = 0 \cdot X_1 + \cdots + 1 \cdot X_k + \cdots + 0 \cdot X_n$ is normal, i.e., all marginal distributions in an n-dimensional normal distribution are one-dimensional normal. However, the reverse is not necessarily true; there are variables $X_1, \ldots, X_n$, each of which is one-dimensional normal, but the vector $(X_1, \ldots, X_n)'$ is not n-dimensional normal.

It is an important consequence of the definition that sums and differences of n-dimensional normal variables have a normal distribution.

If the covariance matrix $\boldsymbol{\Sigma}$ is non-singular, the n-dimensional normal distribution has a probability density function (with $\mathbf{x} = (x_1, \ldots, x_n)'$)

$$
\frac{1}{(2\pi)^{n/2}\sqrt{\det \boldsymbol{\Sigma}}} \exp\left\{ -\frac{1}{2}(\mathbf{x} - \boldsymbol{\mu})'\boldsymbol{\Sigma}^{-1}(\mathbf{x} - \boldsymbol{\mu}) \right\}.
\tag{A.2}
$$

The distribution is said to be *non-singular*. The density (A.2) is constant on every ellipsoid $(\mathbf{x} - \boldsymbol{\mu})'\boldsymbol{\Sigma}^{-1}(\mathbf{x} - \boldsymbol{\mu}) = C$ in $\mathbb{R}^n$.

> **Note:** the density function of an n-dimensional normal distribution is uniquely determined by the expectations and covariances.

Example A.1. Suppose X_1, X_2 have a two-dimensional normal distribution. If

$$\det \Sigma = \sigma_{11}\sigma_{22} - \sigma_{12}^2 > 0,$$

then Σ is non-singular, and

$$\Sigma^{-1} = \frac{1}{\det \Sigma} \begin{pmatrix} \sigma_{22} & -\sigma_{12} \\ -\sigma_{12} & \sigma_{11} \end{pmatrix}.$$

With $Q(x_1, x_2) = (\mathbf{x} - \boldsymbol{\mu})' \Sigma^{-1} (\mathbf{x} - \boldsymbol{\mu})$,

$$Q(x_1, x_2) =$$
$$= \frac{1}{\sigma_{11}\sigma_{22} - \sigma_{12}^2} \left\{ (x_1 - m_1)^2 \sigma_{22} - 2(x_1 - m_1)(x_2 - m_2)\sigma_{12} + (x_2 - m_2)^2 \sigma_{11} \right\} =$$
$$= \frac{1}{1 - \rho^2} \left\{ \left(\frac{x_1 - m_1}{\sqrt{\sigma_{11}}} \right)^2 - 2\rho \left(\frac{x_1 - m_1}{\sqrt{\sigma_{11}}} \right) \left(\frac{x_2 - m_2}{\sqrt{\sigma_{22}}} \right) + \left(\frac{x_2 - m_2}{\sqrt{\sigma_{22}}} \right)^2 \right\},$$

where we also used the correlation coefficient $\rho = \frac{\sigma_{12}}{\sqrt{\sigma_{11}\sigma_{22}}}$, and

$$f_{X_1, X_2}(x_1, x_2) = \frac{1}{2\pi \sqrt{\sigma_{11}\sigma_{22}(1 - \rho^2)}} \exp\left(-\frac{1}{2} Q(x_1, x_2) \right). \qquad (A.3)$$

For variables with $m_1 = m_2 = 0$ and $\sigma_{11} = \sigma_{22} = \sigma^2$, the bivariate density is

$$f_{X_1, X_2}(x_1, x_2) = \frac{1}{2\pi\sigma^2 \sqrt{1 - \rho^2}} \exp\left(-\frac{1}{2\sigma^2(1 - \rho^2)} (x_1^2 - 2\rho x_1 x_2 + x_2^2) \right).$$

We see that, if $\rho = 0$, this is the density of two independent normal variables.

Figure A.1 shows the density function $f_{X_1, X_2}(x_1, x_2)$ and level curves for a bivariate normal distribution with expectations $\boldsymbol{\mu} = (0, 0)$, and covariance matrix

$$\Sigma = \begin{pmatrix} 1 & 0.5 \\ 0.5 & 1 \end{pmatrix}.$$

The correlation coefficient is $\rho = 0.5$. ▲

Remark A.1. *If the covariance matrix Σ is singular and non-invertible, then there exists at least one set of constants $a_1, \ldots, a_n$, not all equal to 0, such that $\mathbf{a}' \Sigma \mathbf{a} = 0$. From (A.1) it follows that $V[\mathbf{a}'\mathbf{X}] = 0$, which means that $\mathbf{a}'\mathbf{X}$ is constant equal to $\mathbf{a}'\boldsymbol{\mu}$. The distribution of $\mathbf{X}$ is concentrated to a hyper plane $\mathbf{a}'\mathbf{x} = $ constant in $\mathbb{R}^n$. The distribution is said to be singular and it has no density function in $\mathbb{R}^n$.*

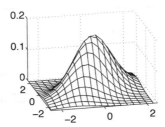 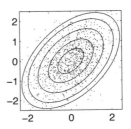

Figure A.1 *Two-dimensional normal density . Left: density function; Right: elliptic level curves at levels* $0.01, 0.02, 0.05, 0.1, 0.15$.

Remark A.2. *Formula (A.1) implies that every covariance matrix Σ is positive definite or positive semi-definite, i.e., $\sum_{j,k} a_j a_k \sigma_{jk} \geq 0$ for all $a_1, \ldots, a_n$. Conversely, if Σ is a symmetric, positive definite matrix of size $n \times n$, i.e., if $\sum_{j,k} a_j a_k \sigma_{jk} > 0$ for all $a_1, \ldots, a_n \neq 0, \ldots, 0$, then (A.2) defines the density function for an n-dimensional normal distribution with expectations m_k and covariances σ_{jk}. Every symmetric, positive definite matrix is a covariance matrix for a non-singular distribution.*

Furthermore, for every symmetric, positive semi-definite matrix, Σ, i.e., such that

$$\sum_{j,k} a_j a_k \sigma_{jk} \geq 0$$

for all $a_1, \ldots, a_n$ with equality holding for some choice of $a_1, \ldots, a_n \neq 0, \ldots, 0$, there exists an n-dimensional normal distribution that has Σ as its covariance matrix.

For n-dimensional normal variables, "uncorrelated" and "independent" are equivalent.

Theorem A.1. *If the random variables $X_1, \ldots, X_n$ are n-dimensional normal and uncorrelated, then they are independent.*

Proof. We show the theorem only for non-singular variables with density function. It is true also for singular normal variables.

If $X_1, \ldots, X_n$ are uncorrelated, $\sigma_{jk} = 0$ for $j \neq k$, then Σ and also Σ^{-1} are diagonal matrices, i.e. (note that $\sigma_{jj} = V[X_j]$),

$$\det \boldsymbol{\Sigma} = \prod_j \sigma_{jj}, \qquad \boldsymbol{\Sigma}^{-1} = \begin{pmatrix} \sigma_{11}^{-1} & \cdots & 0 \\ \vdots & \ddots & \vdots \\ 0 & \cdots & \sigma_{nn}^{-1} \end{pmatrix}.$$

This means that $(\mathbf{x} - \boldsymbol{\mu})' \boldsymbol{\Sigma}^{-1} (\mathbf{x} - \boldsymbol{\mu}) = \sum_j (x_j - \mu_j)^2 / \sigma_{jj}$, and the density (A.2) is

$$\prod_j \frac{1}{\sqrt{2\pi\sigma_{jj}}} \exp \left\{ -\frac{(x_j - m_j)^2}{2\sigma_{jj}} \right\}.$$

Hence, the joint density function for $X_1, \ldots, X_n$ is a product of the marginal densities, which says that the variables are independent. ∎

A.2.1 Conditional normal distribution

This section deals with partial observations in a multivariate normal distribution. It is a special property of this distribution that conditioning on observed values of a subset of variables leads to a conditional distribution for the unobserved variables that is also normal. Furthermore, the expectation in the conditional distribution is linear in the observations, and the covariance matrix does not depend on the observed values. This property is particularly useful in prediction of Gaussian time series, as formulated by the *Kalman filter*.

Conditioning in the bivariate normal distribution

Let X and Y have a bivariate normal distribution with expectations m_X and m_Y, variances σ_X^2 and σ_Y^2, respectively, and with correlation coefficient $\rho = C[X,Y]/(\sigma_X \sigma_Y)$. The simultaneous density function is given by (A.3).

The conditional density function for X given that $Y = y$ is

$$f_{X|Y=y}(x) = \frac{f_{X,Y}(x,y)}{f_Y(y)}$$

$$= \frac{1}{\sigma_X \sqrt{1-\rho^2}\sqrt{2\pi}} \exp \left\{ -\frac{(x - (m_X + \sigma_X \rho (y - m_Y)/\sigma_Y))^2}{2\sigma_X^2(1-\rho^2)} \right\}.$$

Hence, the conditional distribution of X given $Y = y$ is normal with expectation and variance

$$m_{X|Y=y} = m_X + \sigma_X \rho (y - m_Y)/\sigma_Y, \qquad \sigma_{X|Y=y}^2 = \sigma_X^2(1 - \rho^2).$$

Note: the conditional expectation depends linearly on the observed y-value, and the conditional variance is constant, independent of $Y = y$.

Conditioning in the multivariate normal distribution

Let $\mathbf{X} = (X_1, \ldots, X_n)'$ and $\mathbf{Y} = (Y_1, \ldots, Y_m)'$ be two multivariate normal variables, of size n and m, respectively, such that $\mathbf{Z} = (X_1, \ldots, X_n, Y_1, \ldots, Y_m)'$ is $(n+m)$-dimensional normal. Denote the expectations

$$\mathsf{E}[\mathbf{X}] = m_{\mathbf{X}}, \quad \mathsf{E}[\mathbf{Y}] = m_{\mathbf{Y}},$$

and partition the covariance matrix for $\mathbf{Z}$ (with $\mathbf{\Sigma_{XY}} = \mathbf{\Sigma'_{YX}}$),

$$\mathbf{\Sigma} = \mathrm{Cov}\left[\begin{pmatrix} \mathbf{X} \\ \mathbf{Y} \end{pmatrix}; \begin{pmatrix} \mathbf{X} \\ \mathbf{Y} \end{pmatrix}\right] = \begin{pmatrix} \mathbf{\Sigma_{XX}} & \mathbf{\Sigma_{XY}} \\ \mathbf{\Sigma_{YX}} & \mathbf{\Sigma_{YY}} \end{pmatrix}. \tag{A.4}$$

If the covariance matrix $\mathbf{\Sigma}$ is positive definite, the distribution of $(\mathbf{X}, \mathbf{Y})$ has the density function

$$f_{\mathbf{XY}}(\mathbf{x}, \mathbf{y}) = \frac{1}{(2\pi)^{(m+n)/2}\sqrt{\det \mathbf{\Sigma}}} \, e^{-\frac{1}{2}(\mathbf{x}' - m'_{\mathbf{X}}, \mathbf{y}' - m'_{\mathbf{Y}})\mathbf{\Sigma}^{-1}(\mathbf{x} - m_{\mathbf{X}}, \mathbf{y} - m_{\mathbf{Y}})},$$

while the m-dimensional density of $\mathbf{Y}$ is

$$f_{\mathbf{Y}}(\mathbf{y}) = \frac{1}{(2\pi)^{m/2}\sqrt{\det \mathbf{\Sigma_{YY}}}} \, e^{-\frac{1}{2}(\mathbf{y} - m_{\mathbf{Y}})'\mathbf{\Sigma_{YY}^{-1}}(\mathbf{y} - m_{\mathbf{Y}})}.$$

To find the conditional density of $\mathbf{X}$ given that $\mathbf{Y} = \mathbf{y}$,

$$f_{\mathbf{X|Y}}(\mathbf{x} \mid \mathbf{y}) = \frac{f_{\mathbf{YX}}(\mathbf{y}, \mathbf{x})}{f_{\mathbf{Y}}(\mathbf{y})}, \tag{A.5}$$

we need the following matrix property.

Theorem A.2 ("Matrix inversions lemma"). *Let $\mathbf{B}$ be a $p \times p$-matrix ($p = n + m$):*

$$\mathbf{B} = \begin{pmatrix} \mathbf{B}_{11} & \mathbf{B}_{12} \\ \mathbf{B}_{21} & \mathbf{B}_{22} \end{pmatrix},$$

where the sub-matrices have dimension $n \times n$, $n \times m$, etc. Suppose $\mathbf{B}, \mathbf{B}_{11}, \mathbf{B}_{22}$ are non-singular, and partition the inverse in the same way as $\mathbf{B}$,

$$\mathbf{A} = \mathbf{B}^{-1} = \begin{pmatrix} \mathbf{A}_{11} & \mathbf{A}_{12} \\ \mathbf{A}_{21} & \mathbf{A}_{22} \end{pmatrix}.$$

Then

$$\mathbf{A} = \begin{pmatrix} (\mathbf{B}_{11} - \mathbf{B}_{12}\mathbf{B}_{22}^{-1}\mathbf{B}_{21})^{-1} & -(\mathbf{B}_{11} - \mathbf{B}_{12}\mathbf{B}_{22}^{-1}\mathbf{B}_{21})^{-1}\mathbf{B}_{12}\mathbf{B}_{22}^{-1} \\ -(\mathbf{B}_{22} - \mathbf{B}_{21}\mathbf{B}_{11}^{-1}\mathbf{B}_{12})^{-1}\mathbf{B}_{21}\mathbf{B}_{11}^{-1} & (\mathbf{B}_{22} - \mathbf{B}_{21}\mathbf{B}_{11}^{-1}\mathbf{B}_{12})^{-1} \end{pmatrix}.$$

Proof. For the proof, see a matrix theory textbook, for example, [22]. ■

Theorem A.3 ("Conditional normal distribution"). *The conditional normal distribution for* **X**, *given that* **Y** = **y**, *is n-dimensional normal with expectation and covariance matrix*

$$\mathsf{E}[\mathbf{X} \mid \mathbf{Y} = \mathbf{y}] = m_{\mathbf{X}|\mathbf{Y}=\mathbf{y}} = m_{\mathbf{X}} + \Sigma_{\mathbf{XY}} \Sigma_{\mathbf{YY}}^{-1} (\mathbf{y} - m_{\mathbf{Y}}), \quad (A.6)$$

$$\mathsf{C}[\mathbf{X} \mid \mathbf{Y} = \mathbf{y}] = \Sigma_{\mathbf{XX}|\mathbf{Y}} = \Sigma_{\mathbf{XX}} - \Sigma_{\mathbf{XY}}\Sigma_{\mathbf{YY}}^{-1}\Sigma_{\mathbf{YX}}. \quad (A.7)$$

These formulas are easy to remember: the dimension of the submatrices, for example in the covariance matrix $\Sigma_{\mathbf{XX}|\mathbf{Y}}$, is only possible for the matrix multiplications in the right hand side to be meaningful.

Proof. To simplify calculations, we start with $m_{\mathbf{X}} = m_{\mathbf{Y}} = \mathbf{0}$, and add the expectations afterwards. The conditional distribution of **X** given that **Y** = **y** is, according to (A.5), the ratio between two multivariate normal densities, and hence it is of the form,

$$c \exp\left\{ -\frac{1}{2}(\mathbf{x}',\mathbf{y}')\Sigma^{-1} \begin{pmatrix} \mathbf{x} \\ \mathbf{y} \end{pmatrix} + \frac{1}{2}\mathbf{y}'\Sigma_{\mathbf{YY}}^{-1}\mathbf{y} \right\} = c \exp\{-\frac{1}{2}Q(\mathbf{x},\mathbf{y})\},$$

where c is a normalization constant, independent of **x** and **y**. The matrix Σ can be partitioned as in (A.4), and if we use the matrix inversion lemma, we find that

$$\Sigma^{-1} = \mathbf{A} = \begin{pmatrix} \mathbf{A}_{11} & \mathbf{A}_{12} \\ \mathbf{A}_{21} & \mathbf{A}_{22} \end{pmatrix} \quad (A.8)$$

$$= \begin{pmatrix} (\Sigma_{\mathbf{XX}}-\Sigma_{\mathbf{XY}}\Sigma_{\mathbf{YY}}^{-1}\Sigma_{\mathbf{YX}})^{-1} & -(\Sigma_{\mathbf{XX}}-\Sigma_{\mathbf{XY}}\Sigma_{\mathbf{YY}}^{-1}\Sigma_{\mathbf{YX}})^{-1}\Sigma_{\mathbf{XY}}\Sigma_{\mathbf{YY}}^{-1} \\ -(\Sigma_{\mathbf{YY}}-\Sigma_{\mathbf{YX}}\Sigma_{\mathbf{XX}}^{-1}\Sigma_{\mathbf{XY}})^{-1}\Sigma_{\mathbf{YX}}\Sigma_{\mathbf{XX}}^{-1} & (\Sigma_{\mathbf{YY}}-\Sigma_{\mathbf{YX}}\Sigma_{\mathbf{XX}}^{-1}\Sigma_{\mathbf{XY}})^{-1} \end{pmatrix}.$$

We also see that

$$\begin{aligned} Q(\mathbf{x},\mathbf{y}) &= \mathbf{x}'\mathbf{A}_{11}\mathbf{x}+\mathbf{x}'\mathbf{A}_{12}\mathbf{y}+\mathbf{y}'\mathbf{A}_{21}\mathbf{x}+\mathbf{y}'\mathbf{A}_{22}\mathbf{y}-\mathbf{y}'\Sigma_{\mathbf{YY}}^{-1}\mathbf{y} \\ &= (\mathbf{x}'-\mathbf{y}'\mathbf{C}')\mathbf{A}_{11}(\mathbf{x}-\mathbf{C}\mathbf{y})+\tilde{Q}(\mathbf{y}) \\ &= \mathbf{x}'\mathbf{A}_{11}\mathbf{x}-\mathbf{x}'\mathbf{A}_{11}\mathbf{C}\mathbf{y}-\mathbf{y}'\mathbf{C}'\mathbf{A}_{11}\mathbf{x}+\mathbf{y}'\mathbf{C}'\mathbf{A}_{11}\mathbf{C}\mathbf{y}+\tilde{Q}(\mathbf{y}), \end{aligned}$$

for some matrix **C** and quadratic form $\tilde{Q}(\mathbf{y})$ in **y**.

Here $\mathbf{A}_{11} = \boldsymbol{\Sigma}_{\mathbf{XX}|\mathbf{Y}}^{-1}$, according to (A.7) and (A.8), while we can find $\mathbf{C}$ by solving

$$-\mathbf{A}_{11}\mathbf{C} = \mathbf{A}_{12}, \quad \text{i.e.,} \quad \mathbf{C} = -\mathbf{A}_{11}^{-1}\mathbf{A}_{12} = \boldsymbol{\Sigma}_{\mathbf{XY}}\boldsymbol{\Sigma}_{\mathbf{YY}}^{-1},$$

according to (A.8). This is precisely the matrix in (A.6).

If we reinstall the deleted $m_{\mathbf{X}}$ and $m_{\mathbf{Y}}$, we get the conditional density for $\mathbf{X}$ given $\mathbf{Y} = \mathbf{y}$ to be of the form

$$c \exp\{-\frac{1}{2}Q(\mathbf{x},\mathbf{y})\} = c \exp\{-\frac{1}{2}(\mathbf{x}' - m_{\mathbf{X}|\mathbf{Y}=\mathbf{y}}')\boldsymbol{\Sigma}_{\mathbf{XX}|\mathbf{Y}}^{-1}(\mathbf{x} - m_{\mathbf{X}|\mathbf{Y}=\mathbf{y}})\},$$

which is the normal density we were looking for. ■

A.2.2 Complex normal variables

In most of the book, we have assumed all random variables to be real valued. In many applications, and also in the mathematical background, it is advantageous to consider complex variables, simply defined as $Z = X + iY$, where X and Y have a bivariate distribution. The mean value of a complex random variable is simply

$$\mathsf{E}[Z] = \mathsf{E}[\Re Z] + i\mathsf{E}[\Im Z],$$

while the variance and covariances are defined with complex conjugate on the second variable,

$$\mathsf{C}[Z_1,Z_2] = \mathsf{E}[Z_1\overline{Z_2}] - m_{Z_1}\overline{m_{Z_2}},$$
$$\mathsf{V}[Z] = \mathsf{C}[Z,Z] = \mathsf{E}[|Z|^2] - |m_Z|^2.$$

Note, that for a complex $Z = X + iY$, with $V[X] = V[Y] = \sigma^2$,

$$\mathsf{C}[Z,Z] = \mathsf{V}[X] + \mathsf{V}[Y] = 2\sigma^2,$$
$$\mathsf{C}[Z,\overline{Z}] = \mathsf{V}[X] - \mathsf{V}[Y] + 2i\mathsf{C}[X,Y] = 2i\mathsf{C}[X,Y].$$

Hence, if the real and imaginary parts are uncorrelated with the same variance, then the complex variable Z is uncorrelated with its own complex conjugate, $\overline{Z}$. Often, one uses the term *orthogonal*, instead of uncorrelated for complex variables.

A.3 Convergence in quadratic mean

A derivative $X'(t)$ and an integral $\int_a^b g(t)X(t)\,dt$ are limits of an approximating difference quotient $h^{-1}(X(t+h) - X(t))$ and a Riemann sum $\sum g(t_i')X(t_i')(t_i - t_{i-1})$, respectively. Similarly, an infinite sum, $\sum_{k=1}^{\infty} A_k \cos(2\pi f_k t + \phi_k)$, is the limit of $\sum_{k=1}^{n} A_k \cos(2\pi f_k t + \phi_k)$. It is therefore important to define convergence so that it can be checked in terms of the statistical distributions. For (weakly) stationary processes, we need a convergence concept that only depends on the first and second order moment functions.

Definition A.2. *A sequence* $\{X_n, n = 1, 2, \ldots\}$ *of random variables with* $E[X_n^2] < \infty$ *converges to the random variable* X *in quadratic mean (q.m.) (notation* $X_n \overset{q.m.}{\to} X$*), when* $n \to \infty$*, if*

$$E[(X_n - X)^2] \to 0.$$

In particular, if X *is a constant,* $X = c$*, then* $X_n \to c$ *in q.m. if, with* $m_n = E[X_n]$*,*

$$E[(X_n - c)^2] = E\left[(X_n - m_n)^2\right] + (m_n - c)^2 \to 0 \quad \text{when} \quad n \to \infty.$$

It is often simple to prove quadratic mean convergence since no probability arguments are needed. Furthermore, one is, under certain conditions, allowed to interchange the order of the limiting operation and taking expectation.

Theorem A.4. *If* $X_n \to X$ *in quadratic mean, then*
(a) $E[X_n] \to E[X]$*,*
(b) $V[X_n] \to V[X]$*,*
(c) $E[X_n Y] \to E[XY]$ *for every* Y *with* $E[Y^2] < \infty$*,*
(d) $E[X_n Y_n] \to E[XY]$ *if* $Y_n \to Y$ *in q.m.*

Proof. We prove (c) and leave the rest to the reader. Schwarz' inequality for random variables states that $(E[UZ])^2 \leq E[U^2]E[Z^2]$, and it directly gives

$$(E[X_n Y] - E[XY])^2 = (E[(X_n - X)Y])^2 \leq E[(X_n - X)^2]E[Y^2] \to 0$$

when $n \to \infty$, since $E[Y^2] < \infty$ and $X_n \overset{q.m.}{\to} X$.   ■

Theorem A.5 (The Cauchy criterion). *A sequence* $\{X_n, n = 1, 2, \ldots\}$ *converges in quadratic mean if and only if*

$$E[(X_m - X_n)^2] \to 0,$$

when m and n go to infinity independently of each other.

Proof. The "only if" part is easy: Assuming q.m. convergence,

$$E[(X_m - X_n)^2] \le 2E[(X_m - X)^2] + 2E[(X_n - X)^2] \to 0.$$

The "if" part is difficult, and we do not give any proof. In essence, the proof consists of a construction of the limit; in mathematical terms, one shall prove that the set of random variables with finite second moment is *complete*.   ■

Theorem A.6 (The Loève criterion). *A sequence* $\{X_n, n = 1, 2, \ldots\}$ *converges in quadratic mean if and only if there is a constant c such that*

$$E[X_m X_n] \to c, \qquad\qquad\qquad (A.9)$$

when m and n go to infinity independently of each other.

Proof. If condition (A.9) in the Loève criterion is satisfied, then all terms in the right hand side of

$$E[(X_m - X_n)^2] = E[X_m X_m] - 2E[X_m X_n] + E[X_n X_n]$$

have the same limit c, and the left hand side tends to zero. Then X_n converges, according to the Cauchy criterion, proving the "if" part. The "only if" part is the same as Theorem A.4(d): if $X_m, X_n \to X$ in quadratic mean, then $E[X_m X_n] \to E[X^2]$, i.e., $E[X_m X_n]$ has a limit, $c = E[X^2]$.   ■

The Cauchy criterion and the Loève criterion are equivalent.

A.4 Some statistical theory

Statistical inference gives us the means to analyze data in order to find a good statistical model that serves our purposes. We will give a brief summary of the most basic concepts of *parameter estimation*. It is then assumed that we have a statistical model for a random variable X, with a distribution that depends on one or more unknown parameters, and we want, based on observations of X, to suggest good values for the parameters. The procedure should perform well under the condition that the model is correct. We remind the reader of some other aspects of statistical inference, like *model selection*, and *model fitting*, where the purpose is to select a model that describes the data well enough to be useful.

Let $\{X(t), t = 1, 2, \ldots\}$ be a stationary sequence, and let $\{x(t), t = 1, 2, \ldots\}$ denote a realization of the sequence. Thus, $x(t)$ is an observation of the random variable $X(t)$. All $X(t)$ have the same distribution (if the sequence is strictly stationary), or have the same expectation and variance/covariance properties (if the sequence is weakly stationary).

Suppose the distribution of the sequence depends on one or more parameters, $\theta_1, \ldots, \theta_p$, and that these are more or less unknown. A *point estimator* $\hat{\theta}_n$ of the parameter θ is simply a function of the n observations $x(1), \ldots, x(n)$,

$$\hat{\theta}_n(x(1), \ldots, x(n)).$$

Note that when we write $\hat{\theta}_n$ we mean, either the value of the estimator for the obtained observations, i.e., $\hat{\theta}_n(x(1), \ldots, x(n))$, or the random variable, $\hat{\theta}_n(X(1), \ldots, X(n))$.

A point estimator $\hat{\theta}_n$ is called an *unbiased estimator* of θ if, for all possible values of θ, it has expectation

$$E[\hat{\theta}_n(X(1), \ldots, X(n))] = \theta.$$

A weaker property is *asymptotically unbiased*, which means that

$$E[\hat{\theta}_n(X(1), \ldots, X(n))] \to \theta,$$

when $n \to \infty$. The *bias* of an estimator is

$$b(\hat{\theta}_n) = E[\hat{\theta}_n(X(1), \ldots, X(n))] - \theta.$$

Theorem A.7. *If $\{X(t), t = 1, 2, \ldots\}$ is weakly stationary with mean m, then*

$$\hat{m}_n = \frac{1}{n} \sum_{t=1}^{n} x(t)$$

is an unbiased estimator of m, regardless of the covariance function $r_X(\tau)$.

Proof. The expectation of a sum of random variables is equal to the sum of the expectations; this is true both for dependent and independent variables. Therefore

$$E[\hat{m}_n] = \frac{1}{n} \sum_{t=1}^{n} E[X(t)] = m$$

for all m. ∎

An estimator $\hat{\theta}_n$ is called *consistent* or, more precisely, consistent in quadratic mean, if

$$\hat{\theta}_n \to \theta \quad \text{in quadratic mean, when } n \to \infty,$$

i.e.,

$$E[(\hat{\theta}_n - \theta)^2] \to 0 \quad \text{when } n \to \infty.$$

Chebyshev's inequality then implies that

$$P(|\hat{\theta}_n - \theta| > \varepsilon) \le \frac{E[(\hat{\theta}_n - \theta)^2]}{\varepsilon^2} \to 0,$$

for all $\varepsilon > 0$, which means that the estimator is approximately correct when we have many data points.

Since

$$\begin{aligned} E[(\hat{\theta}_n - \theta)^2] &= E[(\hat{\theta}_n - E[\hat{\theta}_n] + E[\hat{\theta}_n] - \theta)^2] \\ &= E[(\hat{\theta}_n - E[\hat{\theta}_n])^2] + (E[\hat{\theta}_n] - \theta)^2 = V[\hat{\theta}_n] + b(\hat{\theta}_n)^2, \end{aligned}$$

an estimator is consistent if it is asymptotically unbiased, $E[\hat{\theta}_n] \to \theta$, and its variance goes to zero, $V[\hat{\theta}_n] \to 0$.

To every estimator one can attach a quality measure, which quantifies the degree of uncertainty. Often its standard deviation, $D[\hat{\theta}_n] = \sqrt{V[\hat{\theta}_n]}$, is used, or, in case this also is unknown, an estimate of the standard deviation, $d = \widehat{D[\hat{\theta}_n]}$, called the *standard error* of the estimator.

A *confidence interval* is a practical way to specify both the value of the estimate and its uncertainty. A confidence interval for a parameter θ with *confidence level* α is an interval, calculated from the observations $x(1), \ldots, x(n)$, written

$$I_\theta(x(1), \ldots, x(n)),$$

such that it, with the prescribed probability $1 - \alpha$, contains the unknown parameter θ,

$$P(\theta \in I_\theta(X(1), \ldots, X(n))) = 1 - \alpha.$$

The probability of coverage, $1 - \alpha$, is to be interpreted as the long run relative frequency of correct statements, when the experiment is repeated.

Appendix B

Delta functions, generalized functions, and Stieltjes integrals

B.1 Introduction

The (Dirac) delta function is a very useful mathematical invention that, in particular, allows us to combine continuous and discrete statistical distributions in one and the same formalism. It has been very much generalized to the class of *(Schwarz) distributions*, or *generalized functions*. Generalized functions are used throughout Chapter 6 to enable a simple general description of linear filters.

In Sections B.2 – B.4 we describe the simplest properties of delta functions, and use them to write statistical distribution and expectations of random variables in a unified way, without having to separate continuous and discrete distributions. Section B.5 contains the most important part of this appendix, a brief discussion of generalized functions. For a good introduction to delta functions and generalized functions, see [41, Ch. II] or the article in [51].

The final section, Section B.6, indicates one way to make the formal calculations in Sections B.2 – B.4 mathematically rigorous.

B.2 The delta function

Usually, the expectation of a random variable X is defined as

$$E[X] = \begin{cases} \int xf(x)\,dx, \\ \sum xp(x), \end{cases}$$

depending on whether X has a continuous distribution with probability density function $f(x)$ or is discrete with probability function $p(x) = P(X = x)$. More generally, one has

$$E[g(X)] = \begin{cases} \int g(x)f(x)\,dx, & \text{for a continuous variable,} \\ \sum g(x)p(x), & \text{for a discrete variable.} \end{cases}$$

The delta function $\delta_c(x)$ is defined only as an integrand in an integration,

$$\int_a^b f(x)\delta_c(x)\,dx = \begin{cases} f(c), & \text{if } c \in (a,b), \\ 0, & \text{otherwise,} \end{cases}$$

for every function $f(x)$ that is continuous at c. With the help of the delta function, one can define a formal *density function*, also for discrete random variables.

Let X be a discrete random variable, whose possible outcomes, $X = x_k$, have probabilities $p(x_k)$, and put

$$f(x) = \sum p(x_k)\delta_{x_k}(x). \tag{B.1}$$

Then, $f(x)$ works exactly as a probability density for continuous variables. For example,

$$\int g(x)f(x)\,ds = \int g(x)\sum_k p(x_k)\delta_{x_k}(x)\,dx = \sum_k p(x_k)\int g(x)\delta_{x_k}(x)\,dx$$
$$= \sum_k p(x_k)g(x_k),$$

which is the formula for the expectation $E[g(x)]$. Formally, we can therefore write the expectation in the same way for continuous and for discrete variables.

Regardless of whether X is continuous or discrete, we have

$$E[g(x)] = \int g(x)f(x)\,dx,$$

if only the density function is defined according to formula (B.1), including the delta functions.

There is also a connection between the formal derivative of the cumulative distribution function $F(x)$ of a random variable X, and the density funtion, expressed with help of delta funtions. For a continuous random variable, the probability density $f(x) = \frac{d}{dx}F(x)$ is the derivative of the distribution function.

If X is discrete, one can formally define its density function as the distributional derivative of the piecewise constant distribution function. Define

$$H_c(x) = \begin{cases} 1 & \text{for } x \geq c, \\ 0 & \text{for } x < c. \end{cases} \tag{B.2}$$

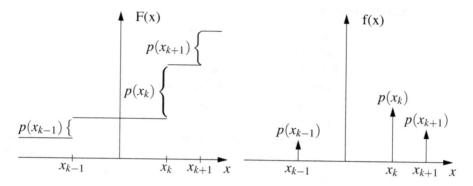

Figure B.1 *Illustration of the formal derivation of a discrete distribution function.*

Then, formally,

$$\frac{dH_c(x)}{dx} = \delta_c(x).$$

For a discrete random variable, we then have $F(x) = \sum_{y \le x} P(y) = \sum_k p(x_k) H_{x_k}(x)$, where x_k are the possible values. We get

$$\frac{dF(x)}{dx} = \sum_k p(x_k) \frac{d}{dx} H_{x_k}(x) = \sum_k p(x_k) \delta_{x_k}(x) = f(x),$$

as required.

B.3   Formal rules for density functions

Many of the well known rules for integration are valid also for integration of delta functions. The first, and most important, rule is that integration is a linear operation; if $f_1(x)$ and $f_2(x)$ are density functions and $g(x)$ an ordinary function, then

$$\int g(x)(c_1 f_1(x) + c_2 f_2(x)) \, dx = c_1 \int g(x) f_1(x) \, dx + c_2 \int g(x) f_2(x) \, dx. \quad \text{(B.3)}$$

One should be aware that it is not enough to assume that $g(x)$ is a density function, because multiplication of delta functions is not defined.

Furthermore, the formula for change of variable in an integral is valid also for delta functions. In particular, the following equality holds,

$$\delta_c(x) = \delta_0(x - c). \quad \text{(B.4)}$$

B.4 Distibutions of mixed type

The standard statistical distributions are either continuous or discrete. By means of the delta function, we can merge the two types into one, and allow distributions to have both a continuous part and a discrete part. A simple example of a variable with such a mixture type distribution is the amount of rain falling during a day. It is either exactly zero, or it is a positive amount, with a continuous distribution.

Suppose a random variable X, with positive probability, takes any of a discrete number of values, x_k, or it takes any value in some interval. Its cumulative distribution function is then of mixed type, consisting of a sum and an integral,

$$F(x) = \int_{-\infty}^{x} \phi(y)\,dy + \sum_{y \leq x} p(y).$$

The derivative, i.e., the probability density, is

$$f(x) = \frac{dF(x)}{dx} = \phi(x) + \sum_{k} p(x_k)\delta_{x_k}(x),$$

and the expectation can be written as an integral,

$$E[g(x)] = \int g(x)f(x)\,dx = \int g(x)\left(\phi(x) + \sum_{k} p(x_k)\delta_{x_k}(x)\right)dx =$$

$$= \int g(x)\phi(x)\,dx + \sum_{k} g(x_k)p(x_k).$$

Example B.1. Calls to a service station will be immediately taken care of, if any service person is available; this happens with probability $p > 0$. Otherwise, and with probability $1 - p$, they have to wait a random time, which we assume to have an exponential distribution with expectation 1. We first find the distribution of the waiting time X, until service begins. Obviously, $F(x) = P(X \leq x) = 0$, for $x < 0$, while, for $x \geq 0$,

$$F(x) = P(X \leq x) = P(X = 0) + P(X \leq x, X > 0) = p + (1 - p)\int_{0}^{x} e^{-y}\,dy.$$

The expectation is, according to the law of total probability,

$$E[X] = pE[X \mid X = 0] + (1 - p)E[X \mid X > 0] = p \cdot 0 + (1 - p) \cdot 1 = 1 - p. \tag{B.5}$$

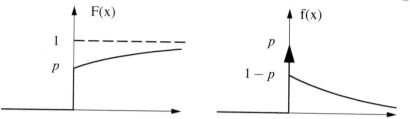

Figure B.2 *Distribution function and probability density for distribution of mixed type.*

The formal derivative of the distribution function is

$$f(x) = \frac{dF(x)}{dx} = p\delta_0(x) + \begin{cases} 0 & \text{if} \quad x < 0, \\ (1-p)e^{-x} & \text{if} \quad x \geq 0; \end{cases}$$

see Figure B.2. Hence,

$$E[X] = \int xf(x)\,dx = \int_{0-}^{0+} xp\,\delta_0(0)\,dx + \int_{0+}^{\infty} (1-p)xe^{-x}\,dx$$
$$= 0 + (1-p) = 1-p,$$

in agreement with (B.5). ▲

B.5 Generalized functions

The calculations above in fact do not require that the functions $f(x)$ are densities, i.e., that they are non-negative and integrate to 1, and in this section we will not make this assumption.

Above we have defined a density as a sum of two components: an ordinary function and a weighted sum of delta functions. This is the most simple form of a generalized function. Mathematically, generalized functions only have meaning as integrands in integrals of "test functions," differentiable functions $g(x)$, which tend to zero suitably fast at $\pm\infty$. Then a generalized function $f(x)$ is defined by partial integration,

$$\int g(x)f(x)\,dx = -\int g'(x)F(x)\,dx,$$

where $F(x) = \int_{-\infty}^{x} f(y)\,dy$. Thus, in particular,

$$\int g(x)\delta_c(x)\,dx = -\int g'(x)H_c(x)\,dx = -\int_c^{\infty} g'(x)\,dx = g(c), \qquad \text{(B.6)}$$

where $H_c(x)$ is defined by (B.2).

Derivatives $\delta'_c(x)$ of delta functions $\delta_c(x)$ are then again defined by partial integration,

$$\int g(x)\delta'_c(x)\,dx = -\int g'(x)\delta_c(x)\,dx = -g'(c), \qquad (B.7)$$

and higher order derivatives are defined recursively, in the same way.

In this text we only use generalized functions that are obtained by adding an ordinary function and a weighted sum of delta functions and derivatives and higher order derivatives of delta functions. Further our use is formal: we allow the input to be any weakly stationary process and define the output by linearity and formulas such as (B.6) and (B.7).

B.6 Stieltjes integrals

The reasoning so far has been "formal," as when we found the "formal derivative" of the function $H_c(x)$ to be $\delta_c(x)$. The mathematical distribution theory gives a precise meaning to these manipulations, but that is outside the scope of this text. Instead, we shall introduce the *Stieltjes*[1] *integral*, which allows us to write integrals as

$$\int g(x)f(x)\,dx = \int g(x)\,dF(x),$$

with $f(x)\,dx = dF(x)$, and $f(x) = dF(x)/dx$ if $F(x)$ is differentiable.

B.6.1 Definition

Let $g(x)$ be a function taking real or complex values on the interval $[a,b]$, and let $F(x)$ be a real valued non-decreasing function, defined on the same interval. Also assume that $F(x)$ is continuous to the right, so that $F(x+0) = \lim_{h\downarrow 0} F(x+h) = F(x)$, and $F(x-0) = \lim_{h\uparrow 0} F(x+h) \leq F(x)$. Then $F(x)$ can have at most a countable number of jump discontinuities in $[a,b]$. Any distribution function $F(x)$ for a random variable X is of the specified type, since it is non-increasing, bounded, and right-continuous at every point.

Now, divide the interval $[a,b]$ into n subintervals by means of the points $a = x_0 < x_1 < \ldots < x_n = b$, and take the sum

$$\sum_{j=1}^{n} g(x'_j)\{F(x_j) - F(x_{j-1})\}, \qquad (B.8)$$

---

[1]Thomas Jan (Joannes) Stieltjes, Dutch mathematician, 1856–1894.

where x'_j is taken anywhere in the interval $[x_{j-1}, x_j]$. If, when $n \to \infty$ and

$$\Delta = \max_j |x_j - x_{j-1}| \to 0,$$

it holds that the sum (B.8) has a limit that is independent of the division, then the limit is defined as the Stieltjes integral of $g(x)$ with regards to $F(x)$, and it is denoted by

$$\int_a^b g(x) \, dF(x).$$

The limit exists, independently of the sub-division, if $g(x)$ is continuous or, more generally, if $g(x)$ has a finite number of discontinuities that do not agree with any discontinuity point of $F(x)$. If $g(x)$ is identically one, the definition directly gives

$$\int_a^b 1 \, dF(x) = F(b) - F(a).$$

To specify the role of the interval endpoints, one should rather write the integral

$$\int_{a+0}^{b+0} g(x) \, dF(x),$$

and also introduce the integral over a single point a, as

$$\int_{a-0}^{a+0} g(x) \, dF(x) = g(a)\{F(a) - F(a-0)\}, \tag{B.9}$$

which is different from zero if $g(a) \neq 0$ and $F(x)$ is discontinuous at $x = a$. Adding or subtracting a term of the type (B.9), we can define the integrals,

$$\int_{a-0}^{b+0} g(x) \, dF(x), \quad \int_{a+0}^{b-0} g(x) \, dF(x), \quad \int_{a-0}^{b-0} g(x) \, dF(x).$$

If $F(x)$ is continuous at both end-points, all these integrals have the same value.

B.6.2 Some simple properties

We shall first consider two special cases, which are important for our applications. If $F(x)$ is continuous and differentiable everywhere, except at finitely many points, then

$$\int_a^b g(x) \, dF(x) = \int_a^b g(x) F'(x) \, dx,$$

which means that the Stieltjes integral can be computed as if it was a common Riemann integral.

If, on the other hand, $F(x)$ is piecewise constant, with jumps with size p_j at the points c_j, $j = 1, \ldots, n$, then the Stieltjes integral is a finite sum,

$$\int_a^b g(x)\, dF(x) = \sum_{j=1}^n g(c_j)\{F(c_j) - F(c_j - 0)\} = \sum_{j=1}^n g(c_j) p_j.$$

One can also define generalized Stieltjes integrals, for example,

$$\int_{-\infty}^\infty g(x)\, dF(x) = \lim \int_a^b g(x)\, dF(x)$$

when $a \to -\infty$, $b \to \infty$, independently of each other, if the limits exist, with the two special cases,

$$\int_{-\infty}^\infty g(x)\, dF(x) = \int_{-\infty}^\infty g(x) F'(x)\, dx,$$

and

$$\int_{-\infty}^\infty g(x)\, dF(x) = \sum_{j=-\infty}^\infty g(c_j)\{F(c_j) - F(c_j - 0)\} = \sum_{j=-\infty}^\infty g(c_j) p_j,$$

respectively, where the number of discontinuity points may be countably infinite.

Example B.2. We started with the desire that $f(x)\, dx$ and $dF(x)$ should mean the same thing. That means that we want the standard rules for integrals to hold, in particular (B.3) and (B.4).

We first prove (B.3), i.e.,

$$\int_a^b g(x)(c_1\, dF_1(x) + c_2\, dF_2(x)) = c_1 \int_a^b g(x)\, dF_1(x) + c_2 \int_a^b g(x)\, dF_2(x).$$

By definition, the left hand side is

$$\int_a^b g(x)(c_1\, dF_1(x) + c_2\, dF_2(x))$$

$$= \lim \sum_{j=1}^n g(x_j')\{(c_1 F_1(x_j) + c_2 F_2(x_j)) - (c_1 F_1(x_{j-1}) + c_2 F_2(x_{j-1}))\}$$

$$= c_1 \lim \sum_{j=1}^n g(x_j')\{F_1(x_j) - F_1(x_{j-1})\} + c_2 \lim \sum_{j=1}^n g(x_j')\{F_2(x_j) - F_2(x_{j-1})\},$$

which is equal to the right hand side.

To prove (B.4), we just need to observe that

$$\int_a^b g(x)\,dH_0(x-c) = \int_a^b g(x)\,dH_c(x) = \begin{cases} g(c) & \text{if } c \in (a,b], \\ 0 & \text{otherwise.} \end{cases}$$

Other standard properties are checked in the same way.   ▲

We finish this section with two general comments about density functions and Stieltjes integrals. First, we have only discussed non-negative densities, i.e., densities $f(x)$, which satisfy $\int_a^b f(x)\,dx \geq 0$, for all $a \leq b$, or, equivalently, we have only discussed the formal derivative of a non-decreasing function $F(x)$. One can also define the formal derivative of the difference between two non-decreasing functions; that "density" can then be negative.

The other comment is more theoretical. We have only encountered distributions with density of the type $f(x) = \phi(x) + \sum a_k \delta_{x_k}(x)$, with an integrable function $\phi(x)$, but this is not the most general form that exists. There are also distributions that are neither continuous, discrete, nor of the mixed type described above.

Appendix C

Kolmogorov's existence theorem

This appendix is more theoretical than the rest of this book, and is not necessary for the understanding of the main text. In Section 1.2, we introduced a stochastic process $\{X(t), t \in T\}$ as a family of random variables defined on a sample space Ω for a statistical experiment. The sample space is an abstract space that is so rich that every possible outcome of the experiment is represented by an element ω in Ω. A stochastic process is then a function $\{X(t, \omega)\}$ of two variables, the outcome ω and the time parameter t. As for random variables, one usually suppresses the argument ω. We did not comment on the subsets A of Ω that define the events of interest, nor did we mention how to define the probability measure that gives probabilities $P(A)$ to the events, and distributions to random variables.

C.1 The probability axioms

Definition C.1. *In the sample space* Ω, *we consider a family of subsets,* $\mathscr{F}$, *that forms a* σ-*algebra, which means that the following holds.*

1. $\Omega \in \mathscr{F}$,
2. $A \in \mathscr{F} \Rightarrow$ *the complement* $\mathscr{A}^* \in \mathscr{F}$,
3. $A_k \in \mathscr{F}$ *for* $k = 1, 2, \ldots \Rightarrow \bigcup_{k=1}^{\infty} A_k \in \mathscr{F}$.

The subsets A in $\mathscr{F}$ *are called* events. *A probability measure* P *on* $(\Omega, \mathscr{F})$ *is a function defined for every event* $A \in \mathscr{F}$, *such that*

a. $0 \leq P(A) \leq 1$ *for every* $A \in \mathscr{F}$,
b. $P(\Omega) = 1$,
c. $P(\bigcup_{k=1}^{\infty} A_k) = \sum_{k=1}^{\infty} P(A_k)$ *if* $A_1, A_2, \ldots$ *are pairwise disjoint.*

The conditions (a–c) are know as the *Kolmogorov axioms* for probabilities. The important news here is that probabilities are only defined for some, but not necessarily all, subsets of Ω.[1]

C.2 Random variables

Now, let X be a random variable, representing the result of a particular measurement in the experiment, i.e., a function defined on Ω, with values $X(\omega); \omega \in \Omega$. The probabilistic interpretation is that if the experiment results in the outcome ω, the numerical value of the measurement is $X(\omega)$. Hence, we expect the following relation to hold:

$$F_X(x) = \mathsf{P}(X \leq x) = \mathsf{P}(\{\omega; X(\omega) \leq x\}), \qquad (\text{C.1})$$

but this obviously requires that the probability function P is well defined for the subset

$$\{\omega; X(\omega) \leq x\},$$

i.e., $\{\omega; X(\omega) \leq x\} \in \mathscr{F}$ for all real x. This is solved by a radical grip: only functions $X(\omega)$ that have this property are allowed to be called random variables!

C.3 Stochastic processes

Now we are ready to construct a stochastic process. Ideally, we would start with

Ω : sample space for a random experiment,
$\mathscr{F}$: family of interesting subsets, i.e., events,
P : probability measure giving probabilites for the events in $\mathscr{F}$.

Then, define

$\{X(t), t \in T\}$: a family or random variables on $(\Omega, \mathscr{F}, \mathsf{P})$,
called a stochastic process,

which finally gives
the ensemble of the process, i.e., the set of possible paths for the process, which are the functions $t \mapsto x(t)$, which can appear as the result of the experiment,

---

[1]A probability measure that gives probabilities to every subset of the sample space Ω must be discrete, i.e., the probabilities are concentrated to a discrete set of elements of the sample space Ω.

and the F-family,

$$F_T : \text{the family of distribution functions.}$$

However, this is not the way one thinks about a stochastic process. One rather starts with the last step in the list, namely the F-family F_T. For example, looking at the Poisson process, Definition 3.2 states that a Poisson process has independent and stationary increments, such that $X(t+h) - X(t)$ is Poisson distributed, with mean λh. By this we have defined all finite-dimensional distributions for $\{X(t), t \geq 0\}$, and hence we has specified the F-family, and nothing more. We did not mention any sample space Ω, σ-algebra $\mathscr{F}$, nor probability measure P.

So, there is one link missing in the theory – does the Poisson process really exist? Does there exist a sample space Ω, σ-algebra $\mathscr{F}$ of subsets, and probability measure P, so we can define a Poisson process? The positive answer is given by the famous Kolmogorov existence theorem from 1933; *Grundbegriffe der Wahrscheinlichkeitsrechnung*, [31]. In this 62-page short book, Kolmogorov codified the mathematical foundations of probability and stochastic processes to the format we are used to today. In the book, he emphasized the special character of *probability*, compared to the general measure theory that was developed in the beginning of the 20th century.

Theorem C.1 (Kolmogorov's existence theorem). *If F_T is a consistent family[a] of distribution functions there exists a sample space Ω, a σ-algebra $\mathscr{F}$, a probability measure P, and a stochastic process $\{X(t), t \in T\}$ defined on $(\Omega, \mathscr{F}, P)$ that has F_T as its family of distribution functions.*

---
[a]The notion "consistent" means that the family F_T satisfies certain simple and obvious conditions, e.g., $F_{t_1,t_2}(x_1,x_2) = F_{t_2,t_1}(x_2,x_1)$ and $F_{t_1,t_2}(x_1,\infty) = F_{t_1}(x_1)$.

Ingredients of the proof: The complete proof goes far beyond the scope of this book, but some ingredients are worth describing. The sample space in the theorem is simply

$$\Omega = \text{the set of all functions from } T \text{ to } \mathbb{R},$$

so every outcome ω in Ω is itself a function $t \mapsto \omega(t)$. A subset of Ω is a set of functions. The definition of the process is therefore simple: the realization is the outcome ω itself, $X(t, \omega) = \omega(t)$.

The definition of the σ-algebra $\mathscr{F}$ and the probability measure P is more complicated, and we content ourselves by noting one has to guarantee that events (subsets) of the form

$$\{X(t_1) \leq x_1, X(t_2) \leq x_2\} = \{\omega; \omega(t_1) \leq x_1, \omega(t_2) \leq x_2)\},$$

and the like, get the probabilities they should have according to the distribution functions in F_T; for this special event $F_{t_1, t_2}(x_1, x_2)$.

The construction in Kolmogorov's existence theorem is therefore the set Ω of all real valued functions $t \in T$, and the selection of certain subsets, which are given probabilities by means of the distribution functions in the family F_t. Examples of sets that get probability are

$$\{x \in \Omega; x(t) \leq u\}, \qquad \text{gets probability } F_t(u),$$
$$\{x \in \Omega; x(t_k) \leq u_k, \text{ for } k = 1, \ldots, n\}, \quad \text{gets probability } F_{t_1, \ldots, t_n}(u_1, \ldots, u_n).$$

If time t is discrete, all interesting events get probability in this way, but when time is continuous, for example $T = \{t \geq 0\}$, many important and interesting events can not be assigned any unique probability by this construction. Important examples of such "exceptional" events are

$$\{x \in \Omega; x(t + h) \to x(t) \text{ as } h \to 0\},$$

and

$$\{x \in \Omega; x \text{ is a continuous function}\}.$$

Kolmogorov's construction therefore does not allow us in general to discuss the probability that the experiment results in a continuous function, nor can one guarantee that the Poisson process only increases in unit jumps. This drawback has to be remedied by other means, which are the topics of more advanced courses in stochastic processes, e.g.,[33]. The classic book by Cramér & Leadbetter (1967), [16], is worth reading. When we talk about the distribution of a stochastic process in this book, we always mean the distributions and probabilities that are defined by the family of finite-dimensional distribution functions.

Appendix D

Covariance/spectral density pairs

$$r(\tau) = \int e^{i2\pi f\tau} R(f)\, df$$

Covariance function $r(\tau)$, $(\alpha > 0)$ continuous time	Spectral density $R(f)$
$e^{-\alpha|\tau|}$	$\dfrac{2\alpha}{\alpha^2+(2\pi f)^2}$
$\dfrac{1}{\alpha^2+\tau^2}$	$\dfrac{\pi}{\alpha} e^{-2\pi\alpha|f|}$
$e^{-\alpha\tau^2}$	$\sqrt{\pi/\alpha}\exp\left(-\dfrac{(2\pi f)^2}{4\alpha}\right)$
$e^{-\alpha|\tau|}\cos 2\pi f_0\tau$	$\dfrac{\alpha}{\alpha^2+(2\pi f_0-2\pi f)^2} + \dfrac{\alpha}{\alpha^2+(2\pi f_0+2\pi f)^2}$
$\begin{array}{ll} 2\alpha & \text{if } \tau=0 \\ 2\frac{\sin 2\pi\alpha\tau}{2\pi\tau} & \text{if } \tau\neq 0 \end{array}$	$\begin{array}{l} 1 \text{ if } |f|\leq\alpha \\ 0 \text{ if } |f|>\alpha \end{array}$
$\begin{array}{ll} 1-\alpha|\tau| & \text{if } |\tau|\leq\frac{1}{\alpha} \\ 0 & \text{if } |\tau|>\frac{1}{\alpha} \end{array}$	$\begin{array}{ll} \frac{1}{\alpha} & \text{if } f=0 \\ \frac{2\alpha}{(2\pi f)^2}\left(1-\cos(\frac{2\pi f}{\alpha})\right) & \text{if } f\neq 0 \end{array}$
discrete time	
$\dfrac{\sigma^2}{1-\theta^2}\theta^{|\tau|}, \quad \tau\in\mathbb{Z}$	$\dfrac{\sigma^2}{1+\theta^2-2\theta\cos 2\pi f}, \quad |f|<1/2$

Appendix E

A historical background

The study of random functions in general, and stationary stochastic processes in particular, can be motivated in many different ways, and there are many different approaches to the theory. One can concentrate on its mathematical aspects, and regard it as a fascinating part of harmonic analysis. Another entrance is via any of the many application areas, which have inspired the theoretical development. In this book, we have tried to present a coherent theory that concentrates on the statistical concepts and aspects, motivated by many different application areas.

Here we present a non-systematic list of highlights from the exchange of ideas between statistical theory and statistical applications. The list ends around the 1950s. Progress after that has been even faster, both in the width and importance of the range of applications, and in the advance of theory. However, this development is still too close to present time for us to even attempt to summarize it.

Models for random phenomena around 1900

The first examples of stochastic models, as they appear in this book, emerged around the turn of the century, 1900. Louis Bachelier,[1] who worked at the Paris Bourse, presented a mathematical analysis of price fluctuations for bonds on a financial market (*Théorie de la spéculation*, 1900), in which the first mathematical model of the Brownian motion can be found. His name is now closely tied to econometrics and statistical finance. Also Albert Einstein's[2] remarkable analysis of the Brownian motion (*Zur Theorie der Brownschen Bewegung*, 1906) is worth mentioning here. The experimental verifications of Einstein's model gave very convincing proofs of the usefulness of stochastic models in the physical world; see Section 5.3.3.

---

[1] Louis Jean-Baptiste Alphonse Bachelier, French mathematician, 1870–1946.
[2] Albert Einstein, German-born theoretical physicist, 1879–1955.

Arthur Schuster[3] worked around the same time on applications of Fourier analysis in physics (*On the investigation of hidden periodicities with application to a supposed 26 day period of meteorological phenomena*, 1897). He coined the term *periodogram* for the empirical squared Fourier coefficients of a time series, and realized the need to investigate its statistical properties before one could conclude on real periodicities. Albert Einstein also worked on a version of the periodogram in a short note from 1914 but that had no impact in practice.

At the time of the mentioned work, probability theory was not a very active research area, and the importance of its contributions to stochastic process theory was not appreciated until much later. Instead, it was the applied fields that stimulated the introduction of statistical concepts, with studies of economic time series and life insurance principles, meteorological data, and noise in electrical devices. It was the transition in the scientific way of thinking (the new "paradigm"), from *deterministic mathematical models* to *statistical models* described by probability laws, that made new progress possible.

Random noise and signal processing

The presentation of the theory of stationary processes that we give in this book has, to a great extent, been shaped by the analysis of electronic devices. The capacity of electronic systems is often limited by the presence of noise, either produced in the device itself, or introduced during the transmission of a signal through an inhomogeneous medium.

An early example of analysis of electronic noise by statistical methods stems from the 1920s, when Harry Nyquist,[4] used Einstein's description of the thermal movements of molecules to compute the noise effect in a resistor. Thereby, he could present a bound for the best achievable measurement precision in an electronic device. After World War II, Claude Shannon[5] extended this general theory, when he showed that it is possible to obtain almost error-free transmission over a noisy channel, if only the flow of information is kept below a certain limit, the channel capacity.

During and after the war, many scientific papers were published on stochastic processes and electronic communication. Practical important studies had been made in the USA by Norbert Wiener[6] on fire control in air de-

---

[3] Sir Franz Arthur Friedrich Schuster, German-French physicist, 1851 – 1934.
[4] Harry Nyquist, Swedish-American engineer, 1889 – 1976.
[5] Claude Elwood Shannon, American engineer and mathematician, 1916 – 2001.
[6] Norbert Wiener, American mathematician and probabilist, 1894 – 1964.

fense systems. Also statistical methods to identify objects in radar echos were developed.

In the Soviet Union, A.N. Kolmogorov[7] had developed similar theories and methods, but these had not been used in practice.

A comprehensive overview with the summary title *Mathematical analysis of random noise*, and with many new results on stationary processes and their role in electrical engineering, was written by Steven O. Rice,[8] and published in two articles 1944–45, [38]. Rice represented a stationary process as a sum of cosines with random amplitudes and phases, just as we often do in this book. The ideas in Rice's work have inspired theoretical work in probability, and also helped to spread the stochastic paradigm to other engineering areas.

A. N. Kolmogoror

Telephone traffic, queues, and computers

It is worth mentioning that the first application of stochastic processes in electrical engineering was motivated by the needs to analyze traffic intensity in telephone networks.

Agner Krarup Erlang[9] worked as a high school teacher in Copenhagen and other places when he in 1904 took up probability theory as a spare time exercise. Via the Danish mathematician Johan Jensen[10] (1859–1925), he was introduced to the managing director of the Copenhagen Telephone Company, and started to work for the company, applying probability theory to telephone traffic. His first paper on the subject (*The theory of probabilities and telephone conversations*) came 1909, and it dealt with the Poisson law for telephone calls.

A. K. Erlang

Conny Palm[11] (*Intensitätsschwankungen im Fernsprechverkehr*, 1943) is one of the founders of *queuing theory*. He introduced Markov-type properties to describe how the future of a stochastic process is influenced by the present ("events with no or limited aftereffects") and he seems to be the first to use the

---

[7]Andrei Nikolaevich Kolmogorov, Russian mathematician and probabilist, 1903 – 1987.

[8]Steven O. Rice, American engineer, 1907 – 1986.

[9]Agner Krarup Erlang, Danish mathematician, 1878 – 1929.

[10]Best known for Jensen's inequality.

[11]Conrad (Conny) Palm, Swedish engineer, mathematician, and probabilist, 1907 – 1951.

term *point process* for a stream of events, like in the Poisson process. Palm was also engaged in the development of early computer technology.

New applications

Around 1950, the successful use of probability and stationary stochastic processes in electrical engineering rapidly spread to many other scientific and engineering fields. We give now some comments on some of these areas, and also briefly describe the development of the theoretical foundations that took place up to around 1950.

Astronomy and meteorology: Astronomy and statistics have a long common history, including the introduction of the method of least squares. The relations declined by the end of the nineteenth century, and during the first half of the twentieth century, physical descriptions dominated astrophysics. An example from cosmology is the observational bias when calculating the distance to different astronomical objects – stars, galaxies, and galaxy clusters. The brighter a star is or the more galaxies a cluster contains, the more

E. L. Scott

likely it is that it is observed, and hence a bias correction is needed. Such statistical corrections have been used in cosmology since the early 1900s.

G. T. Walker

It is notable, however, that at the same time as stochastic processes were introduced in automatic control and oceanography, they were introduced in astronomy, by among others the American astronomer/statistician Elizabeth Scott,[12] who developed stochastic process ideas to use the brightest galaxy in a cluster as a distance indicator. The name *Scott effect* is now used in recognition of her contributions.

Sir Gilbert Thomas Walker[13] worked in India to improve monsoon rain forecasts, and developed the autoregressive process, now a standard model in many areas.

Automatic control and systems theory: In the deterministic control theory that was developed before 1950 there was no room for the irregular and "unpredictable" disturbances that appear in all industrial processes. The theory of feedback control systems was well developed and understood, but in order to construct an optimal regulator, one must have a mathematical model of the disturbances in the system. In the stochastic control theory developed

---

[12]Elizabeth Leonard Scott, American statistician, who started as an astronomer and turned to statistics, (1917 – 1988).

[13]Sir Gilbert Thomas Walker, British physicist, climatologist, and statistician (1868–1958).

during the 1950s and 1960s, the disturbances were described by stationary stochastic processes. It is fair to say that it was the stochastic control theory that made the first manned lunar space flight possible, 1969.

Economy and finance: The Bachelier model for the price of bonds from 1900 modeled the fluctuations as a Gaussian random walk, in the limit a continuous, but very irregular curve. This model did not have any impact on economic research until in the nineteen sixties.

Instead, starting in the nineteen twenties, *time series* models with time counted in days, months, or years were developed by statisticians like Harold Hotelling[14] in the USA, George U. Yule[15] in England, and Herman Wold[16] in Sweden.

G. U. Yule

The Bachelier model re-emerged in the economic literature after 1960, and was refined to the famous *Black and Scholes' formula*. It is now a leading model in stochastic calculus and financial mathematics, often together with discrete time non-linear models.

Oceanography: Lord Rayleigh, famous British physicist, is claimed to have said (1880): "The basic law of the seaway [= ocean waves] is the apparent lack of any law." True, or not, the quotation illustrates the mental difference between the well-developed mathematical, deterministic theory of waves, in particular water waves, and the impression of confusion from a stormy ocean. H. Sverdrup[17] and W. Munk[18] (1942, 1947) pioneered statistical descriptions of wave height, in an attempt to classify the statistical properties of the ocean. The broad exposition by S.O. Rice, mentioned above, inspired M. St Denis[19] and W.J. Pierson[20] to break with the dominating purely deterministic approaches to wind driven ocean waves and ship motion. In their main work, *Motions of ships in confused seas* (1953), one meets most of the concepts of stationary processes that we deal with in the present book, including linear filters and random fields.

Some more application areas: Other areas where stationary stochastic processes have been introduced, and changed the ways of approaching prob-

---

[14]Harold Hotelling, American mathematical statistician, 1895 – 1973.

[15]George Udny Yule, British statistician, 1871 – 1951.

[16]Herman Wold, Swedish statistician, 1908 – 1992.

[17]Harald U. Sverdrup, Norwegian oceanographer, 1888 – 1957.

[18]Walter H. Munk, Austian-born American oceanographer, 1917 – .

[19]Manley St. Denis, Austrian-born American oceanographer, 1910 –2003.

[20]Willard J. Pierson Jr., American oceanographer, 1922 – 2003.

lems, are: *Image analysis*, where an image intensity can be regarded as a stochastic surface; *Structural reliability*, where the environmental loads on a building or other structure are modeled as a stationary process; and *Environmetrics and climatology*, where global circulation models are described as very complex and interdependent stochastic processes.

Data, computers, and algorithms

The spectrum is a main theme in this book and its estimation is central in time series analysis and signal processing.

The statistical properties of the periodogram was well understood around 1950. In practice, analogue methods dominated until the advent of high-speed computers, but the real breakthrough came with the (re)invention of the Fast Fourier Transform, by Cooley and Tukey 1965, [15]. By use of the FFT the smoothed periodogram estimator became the favored method.

J. W. Tukey

The theoretical development

With the growing interest in time variable random phenomena that took place in the beginning of the twentieth century, the need for a solid mathematical foundation became a matter of importance. It was realized that the most promising line to follow was that of mathematical measure theory, but it had to allow for the interpretation of a *random function*, i.e., one should be able to assign probabilities to the trajectories of the process. A.N. Kolmogorov's *Grundbegriffe der Wahrscheinlichkeitsrechnung*, 1933, added the necessary probabilistic ideas to the measure theory, and an intense period of theoretical development followed, a development that is still very much under way.

Substantial contributions were made by, i.a., A.Y. Khinchin[21] in the Soviet Union, Harald Cramér[22] and Ulf Grenander[23] in Sweden, Paul Lévy[24] in

---

[21] Aleksandr Yakovlevich Khinchin, Russian mathematician and probabilist, 1894 – 1959.

[22] Harald Cramér, Swedish probabilist and statistician, 1893 – 1985.

[23] Ulf Grenander, Swedish mathematician, 1923 – .

[24] Paul Lévy, French mathematician and probabilist, 1886 – 1971.

France, John L. Doob[25] and Will Feller[26] in the USA, K. Ito[27] in Japan, and Ted Hannan[28] in Australia.

Acknowledgment

The images in this section are from the MacTutor History of Mathematics web site: http://www-history.mcs.st-and.ac.uk/

---

[25]John Leo Doob, American mathematician and probabilist, 1910 – 2004.
[26]William Feller, Croatian-born probabilist, 1906 – 1970.
[27]Kiyoshi Ito, Japanese mathematician, 1915 – 2008.
[28]Edward (Ted) James Hannan, Australian statistician, 1921 – 1994.

References

[1] P. Andrén (2006): Power spectral density approximations of longitudinal road profiles. *Int. J. Vehicle Design*, **40**, 2–14.

[2] S. Asmussen and P.W. Glynn (2007): *Stochastic Simulation: Algorithms and Analysis*. Springer, New York.

[3] D.L. Bartholomew and J.A. Tague (1995): Quadratic power spectrum estimation with orthogonal frequency division multiple windows. *IEEE Trans. on Signal Processing*, **43**, 1279–1282.

[4] J.S. Bendat and A.G. Piersol (2010): *Random Data*. 4th ed. Wiley, New York.

[5] J.J. Benedetto (1992): Irregular sampling and frames. In *Wavelets: A Tutorial in Theory and Applications*, Ed: C.K. Chui, pp. 445–507. Academic Press, San Diego.

[6] R.B. Blackman and J.W. Tukey (1958): *The Measurements of Power Spectra*. Dover, New York.

[7] P. Bloomfield (2000): *Fourier Analysis of Time Series: An Introduction*. 2nd ed., Wiley, New York.

[8] K. Bogsjö, K. Podgorski, and I. Rychlik (2012): Models for road surface roughness. *Vehicle System Dynamics*, **50**, 725 – 747.

[9] T. Bollerslev (1986): Generalized autoregressive conditional heteroscedasticity. *J. Econometrics*, **31**, 307–327.

[10] P.J. Brockwell and R.A. Davis (2002): *Introduction to Time Series and Forecasting*. 2nd ed., Springer-Verlag, New York.

[11] P.J. Brockwell and R.A. Davis (1991): *Time Series: Theory and Methods*. 2nd ed., Springer-Verlag, New York.

[12] A.F. Carr (1991): *Point Processes and their Statistical Inference*. 2nd ed., Marcel Dekker, New York.

[13] S. Chandrasekhar (1943): Stochastic problems in physics and astronomy. *Reviews of Modern Physics*, **15**, 1–89. Reprinted in Wax, N.

(1954): *Selected Papers in Noise and Stochastic Processes*. Dover Publications, New York.

[14] M.P. Clark and C.T. Mullis (1993): Quadratic estimation of the power spectrum using orthogonal time-division multiple windows. *IEEE Trans. on Signal Processing*, **41**, 222–231.

[15] J.W. Cooley and J.W. Tukey (1965): An algorithm for the machine calculation of Fourier series. *Math. Comput.*, **19**, 297–301.

[16] H. Cramér and M.R. Leadbetter (1967): *Stationary and Related Stochastic Processes*. John Wiley & Sons, New York. Reprinted by Dover Publications, 2004.

[17] D.J. Daley and D. Vere-Jones (2003): *An Introduction to the Theory of Point Processes*. Vol I, 2nd ed., Springer, New York.

[18] A. Einstein (1906): Zur Theorie der Brownschen Bewegung. *Annalen der Physik*, **19**, 371–381.

[19] P. Embrechts and M. Maejima (2002): *Selfsimilar processes*. Princeton University Press, Princeton.

[20] R.F. Engle (1982): Autoregressive conditional heteroscedasticity with estimates of the variance of United Kingdom inflation. *Econometrica*, **50**, 987–1008.

[21] C. Gallet and C. Julien (2011): The significance threshold for coherence when using the Welch's periodogram method: Effect of overlapping segments. *Biomedical Signal Processing and Control*, **6**, 405–409.

[22] F.A. Graybill (1969): *Introduction to Matrices with Applications in Statistics*, Wadsworth, Belmont.

[23] U. Grenander (1950): Stochastic processes and statistical inference. *Ark. Mat.*, **1**, 195–277.

[24] G. Grimmett and D. Stirzaker (2001): *Probability and Random Processes*. 3rd ed., Oxford University Press, Oxford.

[25] A. Gut (2009): *An Intermediate Course in Probability*, 2nd ed., Springer, Dordrecht.

[26] K. Hasselmann et al. (1973): Measurements of wind-wave growth and swell decay during the JOint North Sea Wave Project (Jonswap). *Deutsche Hydrographische Zeitschrift, Reihe A*, **8**.

[27] L. Isserlis (1918): On a formula for the product-moment correlation of any order of a normal frequency distribution in any number of variables, *Biometrika*, **12**, 134–139.

[28] R.E. Kalman (1960): A new approach to linear filtering and predictor problems. *Trans. ASME-J. Basic Engineering, Ser. D*, **82**, 35–45.

[29] R.W. Katz (2002): Sir Gilbert Walker and a connection between El Niño and statistics, *Statistical Science*, **17**, 97–112.

[30] S.M. Kay (1988): *Modern Spectral Estimation: Theory and Application.* Prentice-Hall, Englewood Cliffs.

[31] A.N. Kolmogorov (1933): *Grundbegriffe der Wahrscheinlichkeitsrechnung.* Springer-Verlag, Berlin.

[32] A.N. Kolmogorov (1956): *Foundations of the Theory of Probability*, 2nd English Edition, Chelsea Publishing Company, New York, (Translation of [31].)

[33] G. Lindgren (2012): *Stationary Stochastic Processes: Theory and Applications.* Chapman & Hall/CRC, Boca Raton.

[34] R.M. Mazo (2002): *Brownian motion. Fluctuations, Dynamics and Application.* Oxford University Press, Oxford.

[35] A. Moberg, H. Tuomenvirta, and Ø. Nordli (2005): Recent climatic trends. In *Physical Geography of Fennoscandia.* Ed: M. Seppälä. Oxford Regional Environments Series, Oxford University Press, Oxford, 113–133.

[36] D.B. Percival and A.T. Walden (1993): *Spectral Analysis for Physical Applications: Multitaper and Conventional Univariate Techniques.* Cambridge University Press, Cambridge.

[37] D.S.G. Pollock (1999): *A handbook of Time-series Analysis, Signal Processing and Dynamics.* Academic Press, New York.

[38] S.O. Rice (1944 & 1945): Mathematical analysis of random noise. *Bell System Technical Journal*, **23**, 282–332 and **24**, 46–156. Reprinted in Wax, N. (1954): *Selected Papers in Noise and Stochastic Processes.* Dover Publications, New York.

[39] K.S. Riedel (1995): Minimum bias multiple taper spectral estimation. *IEEE Trans. on Signal Processing*, **43**, 188–195.

[40] A. Schuster (1898): On the investigation of hidden periodicities with application to a supposed 26 day period of meterological phenomena. *Terr. Magnet.*, **3**, 13–41.

[41] L. Schwartz (2008): *Mathematics for the Physical Sciences.* Dover Publications, New York.

[42] D. Slepian (1978): Prolate spheriodal wave functions, Fourier analysis and uncertainty-v: The discrete case. *Bell System Journal*, **57**, 1371–1430.

[43] M. von Smoluchowski (1906): Zur kinetischen Theorie der Brownschen Molekularbewegung und der Suspensionen. *Annalen der Physik*, **21**, 756–780.

[44] P. Stoica and R. Moses (2005): *Spectral Analysis of Signals*. Pearson, Prentice Hall, NJ.

[45] H.U. Sverdrup and W.H. Munk (1947): Wind, sea, and swell: Theory of Relations for Forecasting. *U.S. Navy Hydrographic Office, H.O.*, Publ. No. 601.

[46] D.J. Thomson (1982): Spectrum estimation and harmonic analysis. *Proc. of the IEEE*, **70**, 1055–1096.

[47] R.S. Tsay (2010): *Analysis of Financial Time Series*, 3rd ed. Wiley, Hoboken.

[48] The WAFO group (2011): WAFO – a MATLAB Toolbox for Analysis of Random Waves and Loads; Tutorial for WAFO version 2.5. Available at http://www.maths.lth.se/matstat/wafo/.

[49] A.T. Walden, E. McCoy, and D.B. Percival (1994): The variance of multitaper spectrum estimates for real Gaussian processes. *IEEE Trans. on Signal Processing*, **42**, 479–482.

[50] G.T. Walker (1931): On periodicities in series of related terms. *Proc. R. Soc. Lond. A*, **131**, 518–532.

[51] http://en.wikipedia.org/wiki/Distribution_(mathematics)

[52] P.D. Welch (1997): The use of fast Fourier transform for the estimation of power spectra: A method based on time averaging over short, modified periodograms. *IEEE Trans. on Audio Electroacoustics*, **AU-15**, 70–73.

[53] C.W. Wright et al. (2001): Hurricane directional wave spectrum spatial variation in the open sea, *J. Phys. Oceanogr.*, **31**, 2472.

[54] A.M. Yaglom (1987): *Correlation Theory of Stationary and Related Random Functions, I-II*. Springer-Verlag, New York.

[55] A.M. Yaglom (1962, 2004): *An Introduction to the Theory of Stationary Random Functions*. Reprint of 1962 edition. Dover Publications, New York.

[56] G.U. Yule (1927): On a method of investigating periodicities in distributed series, with special reference to Wolfer's sunspot numbers. *Phil. Trans. R. Soc., A*, **226**, 267–298.

[57] L. Zetterqvist (1989): *Statistical Methods for Analysing Water Quality Data*. Mathematical statistics, Lund University.

[58] K.J. Åström (1970): *Introduction to Stochastic Control*. Academic Press, New York.

[59] K.J. Åström and C.G. Källström (1976): Identification of ship steering dynamics. *Automatica*, **12**, 9–22.

[60] B. Øksendal (2003): *Stochastic Differential Equations, an Introduction with Applications*. 6th ed., Springer, New York.

Index

CPSIA information can be obtained
at www.ICGtesting.com
Printed in the USA
BVHW040907241118
533266BV00009B/175/P

9 781466 586185